Chemical Dynamics in Fresh Water Ecosystems

Edited by

Frank A.P.C. Gobas, B.Sc., M.Sc., Ph.D.
Assistant Professor
School of Resource-Environmental Management
Simon Fraser University
Burnaby, British Columbia, Canada

John A. McCorquodale, B.E.Sc., M.Sc., Ph.D.
Professor
Department of Civil-Environmental Engineering
University of Windsor
Windsor, Ontario, Canada

CRC Press
Taylor & Francis Group
Boca Raton London New York

CRC Press is an imprint of the
Taylor & Francis Group, an **informa** business

First published 1992 by CRC Press
Taylor & Francis Group
6000 Broken Sound Parkway NW, Suite 300
Boca Raton, FL 33487-2742

Reissued 2018 by CRC Press

Library of Congress Cataloging-in-Publication Data

Chemical dynamics in fresh water ecosystems / edited by Frank A. P. C. Gobas.
    p. cm.
  Includes bibliographical references and index.
  ISBN 0-87371-511-X
  1. Water quality. 2. Freshwater ecology. 3. Water chemistry.
I. Gobas, Frank A. P. C. II. McCorquodale, John A. (John Alex)
TD370.C49 1992
628.1'68—dc20   92-12516

A Library of Congress record exists under LC control number: 92012516

Publisher's Note
The publisher has gone to great lengths to ensure the quality of this reprint but points out that some imperfections in the original copies may be apparent.

Disclaimer
The publisher has made every effort to trace copyright holders and welcomes correspondence from those they have been unable to contact.

ISBN 13: 978-1-315-89142-2 (hbk)
ISBN 13: 978-1-351-07052-2 (ebk)

Visit the Taylor & Francis Web site at http://www.taylorandfrancis.com and the
CRC Press Web site at http://www.crcpress.com

Frank Gobas is an Assistant Professor in the School of Resource and Environmental Management at Simon Fraser University in Burnaby, Vancouver, British Columbia. He received a B.Sc. from the Free University of Amsterdam, an M.Sc. in Environmental and Toxicological Chemistry from the University of Amsterdam and a Ph.D. in Chemical Engineering and Applied Chemistry from the University of Toronto.

His research is focused on the behavior and effects of toxic substances in the environment and the application of environmental modeling as a tool for the management of contaminants in the environment. His work includes studies of the uptake and bioaccumulation of various organic substances in fish, plants, and other organisms, modeling the dynamics of chemical distribution in aquatic ecosystem and food-chains, and relationships between chemical structure and environmental fate of toxic organics.

Dr. Gobas served as a consultant to the International Joint Commission, the Ontario Ministry of the Environment, the Ministry of the Environment, Land and Parks of British Columbia, and Environment Canada on several occasions.

John A. McCorquodale is a professor of Civil and Environmental Engineering at theUniversity of Windsor, Windsor, Ontario, Canada. He holds a B.E.Sc. from the University of Western Ontario, an M.Sc. in Fluid Mechanics from the University of Glasgow, and a Ph.D. in Hydraulic Engineering from the University of Windsor.

His research interests include physical and numerical modeling of hydrodynamic systems. He has developed fate and transport models for large river systems. These models have been used for the simulation of exposure in the water and sediment in the connecting channels of the Great Lakes. Both near-field and far-field concerns have been considered.

Dr. McCorquodale has been involved in contract research with the Ontario Ministry of the Environment; Environment Canada; the Ontario Ministry of Transportation; several municipalities including the City of Windsor, Metropolitan Seattle, and the City of Portland; as well as a number of engineering consulting firms. He has served six years on research grant selection panels of the Natural Sciences and Engineering Research Council of Canada. He is presently a member of the Technological Committee of the International Joint Commission.

PREFACE

The purpose of this book is to bring together the scientific knowledge that environmental engineers, managers, and decision makers need to predict the impact of chemical discharges in aquatic ecosystems. In essence, this book compiles the expertise to establish the relationship between a chemical emission and the resulting ambient concentrations in water, sediments, fish, benthos, plants, and other components of real aquatic ecosystems. The construction of these emission-impact relationships requires scientific expertise from various disciplines, including hydrodynamics, sediment dynamics, chemical fate processes, bioaccumulation, food chain transfer, and others. The chapters in this book summarize the present state of knowledge in these disciplines and discuss models that can be used to make quantitative predictions of the chemical dynamics and behavior in the aquatic environment.

The need for the integration of scientific expertise to determine the relationships between chemical emissions and environmental concentrations and impacts follows from a simple analysis of some of the largest environmental problems of our times. Problems such as ozone depletion, acid rain, global warming, eutrophication, and toxic substances in the food chain (e.g., DDT, PCB, dioxins, mercury, and others) are the result of a poor or complete lack of management and control over chemical emissions into the environment. Production and subsequent release of chemical substances like chlorofluorocarbons, sulphur and nitrogen oxides, carbon dioxide, phosphates, DDT, PCBs, dioxins, and many others in the environment have largely occurred in ignorance of possible impacts on the environment. A good example is polychlorinated biphenyls (PCBs), which were produced for many years and in such large quantities that they can now be found all over our planet. The impact of PCB contamination is particularly evident in the Great Lakes Basin, where the reproductive ability of various bird species is impaired, commercial fishing is curtailed because fish are no longer considered to be fit for consumption, and concentrations of PCB in human breast milk have climbed to levels that are now of serious concern. It was only after the costly ecological and economic damage was done, that strategies for reducing further emissions were put in place.

It is clear that with the current production of more than 50,000 organic substances and the production of approximately 1,000 new chemicals each year, we can no longer rely on ignorance to ensure our standards of environmental quality and to prevent economic losses for our resource and other industries. What is needed are reliable tools that can help the environmental manager and decision maker to determine the impacts of chemical substances in the environment before the chemical substance is discharged, or preferably before the chemical substance is manufactured. Construction of these tools requires a multidisciplinary approach combining the expertise from disciplines which traditionally have had little common ground. It is the objective of this book to provide an integration of sciences for the purpose of sound environmental management.

The chapters in this book draw from areas such as hydrodynamics, sediment dynamics, air-water transport, bioaccumulation, food chain transfer, toxicology, and environmental chemistry. Each chapter summarizes the present state of knowledge in those areas with an emphasis on those processes and phenomena controlling the dynamics and behavior of chemical substances in aquatic ecosystems.

Chapter 1 reviews the principles of hydrodynamics of rivers, embayments in rivers, and shallow lakes. The emphasis of this chapter is on the hydrodynamics of large rivers and the relationship between hydrodynamics and chemical fate. The governing equations and numerical models for their solution are presented with examples of their practical application in the connecting channels of the Great Lakes.

Chapter 2 is a review of the application of numerical hydrodynamic models to the Great Lakes. The main purpose of this chapter is to categorize the various types of hydrodynamic models according to physical processes being modeled and to evaluate the application of hydrodynamic models in solving environmental problems in the Great Lakes.

Chapter 3 focuses on sediment dynamics. This chapter reviews the background terminology, formulation, and adequacy of the parameterization of the sediment exchange process and evaluates the performance of various methods under field conditions. Laboratory and theoretical parameterizations which are based on equilibrium assumptions are discussed in the context of field circumstances which differ from conditions that approach equilibrium.

Chapter 4 discusses the exchange of chemicals between lakes and the atmosphere. This chapter reviews this issue by first presenting a qualitative description of the exchange processes and discussing the factors that influence the rates of these processes. This is followed by a more detailed review and the development of quantitative expressions describing these processes and the individual partitioning phenomena. Finally, these expressions are combined to describe the overall air-lake cycling phenomena.

In Chapter 5, the uptake kinetics and bioaccumulation of polyaromatic hydrocarbons (PAHs) in the benthic amphipod, *Diporeia*, is discussed. This chapter first reviews studies of the physiological and environmental factors controlling the uptake and elimination kinetics of PAHs in *Diporeia*. Then, it presents a pharmacokinetic model for estimating PAH concentrations in the amphipods at any point during the year, which is compared to field data. Finally, this chapter discusses the role of water and sediments as sources for chemical uptake in the amphipods.

Chapter 6 reviews the uptake and bioaccumulation of organic chemicals from the water and sediments into single aquatic organisms, such as fish, plankton, benthic invertebrates, and entire food chains. It summarizes the current state of knowledge regarding the mechanism of chemical uptake and bioaccumulation in various aquatic organisms and presents a model to predict the accumulation of organic substances in aquatic food chains. It is also shown how this model can be used to assess chemical toxicity in fish and other aquatic organisms. The ability of the food chain model to predict chemical concentrations in aquatic food chains is demonstrated by the application of the model to experimental food chain data for Lake Ontario.

Chapter 7 reviews models of organic chemical uptake and transfer in food webs. Two general classes of food chain models are discussed, i.e., (1) generic equilibrium models and (2) site specific models which may be time- and age-dependent. This chapter provides a detailed analysis of how the various food chain models compare in their ability to predict chemical concentrations in organisms of aquatic food chains.

Chapter 8 is a case study of the combination of laboratory studies, mesocosm studies in the field and the application of food chain models to determine the relationship between the emission of chlorinated dioxins and dibenzofurans in pulp mill effluents, and the concentrations of these substances in fish and benthic invertebrates of riverine food chains. The case study is based on field data from various river systems in Canada and applies the food chain models discussed in Chapters 6 and 7.

Chapter 9 is another case study, focusing on the application of chemical fate models and statistical techniques to determine the temporal trends and distribution of PCB congeners in a small isolated lake in eastern Ontario. The purpose of this chapter is to illustrate the present state of knowledge regarding chemical dynamics in real environmental systems by focusing on a small, relatively well-controlled and characterized lake system.

Frank A.P.C. Gobas, Ph.D.
John A. McCorquodale, Ph.D.

CONTRIBUTORS

Gordon C. Balch, B.Sc.
Research Assistant
Environmental Resource Studies
  Program
Trent University
Peterborough, Ontario, Canada

Keith W. Bedford, Ph.D.
Professor
Department of Civil Engineering
The Ohio State University
Columbus, Ohio

John P. Connolly, Ph.D.
Professor
Department of Environmental
  Engineering-Science
Manhattan College
Riverdale, New York

Brian J. Eadie, Ph.D.
Research Assistant
NOAA Great Lakes Environmental
  Research Laboratory
Ann Arbor, Michigan

Wayne L. Fairchild, Ph.D.
Research Assistant
Pesticide Research Laboratory
Department of Soil Science
University of Manitoba
Winnipeg, Manitoba, Canada

Warren R. Faust, Ph.D.
Research Assistant
NOAA Great Lakes Environmental
  Research Laboratory
Ann Arbor, Michigan

Thomas D. Fontaine, Ph.D.
Research Assistant
NOAA Great Lakes Environmental
  Research Laboratory
Ann Arbor, Michigan

Frank A. P. C. Gobas, Ph.D.
Assistant Professor
School of Resource-Environmental
  Management
Simon Fraser University
Burnaby, British Columbia, Canada

Peter F. Landrum, Ph.D.
Research Assistant
NOAA Great Lakes Environmental
  Research Laboratory
Ann Arbor, Michigan

Gregory A. Lang, Ph.D.
Research Assistant
NOAA Great Lakes Environmental
  Research Laboratory
Ann Arbor, Michigan

Colin R. MacDonald, Ph.D.
Research Assistant
AECL Research
Environmental Sciences Branch
Whiteshell Laboratories
Pinawa, Manitoba, Canada

Donald MacKay, Ph.D.
Professor
Institute of Environmental Studies
University of Toronto
Toronto, Ontario, Canada

John A. McCorquodale, Ph.D.
Professor
Department of Civil-Environmental
 Engineering
University of Windsor
Windsor, Ontario, Canada

Chris D. Metcalfe, Ph.D.
Professor
Environmental Resource Studies
 Program
Trent University
Peterborough, Ontario, Canada

Tracy Metcalfe, B. Sc.
Research Scientist
Environmental Resource Studies
 Program
Trent University
Peterborough, Ontario, Canada

Derek C. G. Muir, Ph.D.
Research Scientist
Department of Fisheries and Oceans
Freshwater Institute
Winnipeg, Manitoba, Canada

Thomas Parkerton, M.S., Ph.D.
Research Assistant
Department of Environmental
 Engineering-Science
Manhattan College
Riverdale, New York

David J. Schwab, Ph.D.
Research Assistant
NOAA Great Lakes Environmental
 Research Laboratory
Ann Arbor, Michigan

Robert V. Thomann, Ph.D.
Professor
Department of Environmental
 Engineering-Science
Manhattan College
Riverdale, New York

Mike D. Whittle, Ph.D.
Research Scientist
Department of Fisheries and Oceans
Bayfield Institute
Burlington, Ontario, Canada

Alvin L. Yarechewski
Research Assistant
Department of Fisheries and Oceans
Freshwater Institute
Winnipeg, Manitoba, Canada

Ee-Mun Yuen, Ph.D.
Associate Professor
Lawrence Technological University
Southfield, Michigan

CONTENTS

1 HYDRODYNAMICS OF CONNECTING CHANNELS

INTRODUCTION

The natural system of connecting channels in the Great Lakes is heavily used for water supply, wastewater disposal, and transportation. Some of these activities have resulted in a degradation of the water quality in and downstream of the connecting channel. The importance of maintaining or restoring the environmental quality of these connecting channels resulted in joint research effort by Canadian and U.S. researchers, known as the Upper Great Lakes Connecting Channels Study (UGLCC 1988). The UGLCC study lasted three years and culminated in the development of both mass balance and process oriented models. These models are intended to be used as tools for research planning and for resource management in the connecting waterways of the Great Lakes.

Recently the Great Lakes Water Quality Agreement between Canada and the U. S. was renewed. The explicitly stated goal of the new Canada-U.S. agreement is the virtual elimination of persistent toxics from discharges to the Great Lakes. Shortly after this agreement, the Ontario Ministry of Environment (MOE) adopted the Municipal-Industrial Strategy for Abatement (MISA, 1986) plan. MISA provides a mechanism for continuous reductions in municipal and industrial water pollution of the Great Lakes. All major industrial and municipal dischargers to Ontario waterways in the Great Lakes will be subject to monitoring and regulation which require them to report the concentrations and total amounts of a broad range of contaminants in their effluents. This self-monitoring program will be audited by the MOE.

Information from the monitoring will then be used to formulate an abatement regulation that will specify allowable concentrations, as well as loadings, of toxic pollutants for each discharger. Well-calibrated and verified mathematical models are required at this stage to estimate the wasteload allocation to meet specified water quality objectives in the receiving waters. The discharge limits will then be compared with loads achievable using the best available technology that is

ISBN 0-87371-511-X

economically achievable (BAT-EA). The ultimate goal is to achieve the lower of these two limits.

MODELING OF CONNECTING CHANNELS

The first stage of developing a connecting channel model is to build and test a hydrodynamic submodel. This provides the advective and turbulent mass transfer coefficients for the second stage models which are the chemical fate and transport submodels. Together these submodels simulate the flow patterns and exposure levels in the aquatic ecosystem of the connecting channels. The hydrodynamic and mass transport models must have the capability of simulating the range of expected flow conditions.

The third stage in the modeling exercise is to use the resulting output from the hydrodynamic and transport models to drive the uptake of toxic pollutants into the aquatic food web. The ultimate objective of an applied modeling study of the connecting channels is to predict future water quality impacts resulting from changes in management strategies under an expected or observed range of natural conditions. The predictive capability of these site-specific models will allow the environmental managers to screen the myriad of management alternatives when preparing preliminary waste load allocations for both conventional and toxic pollutants.

HYDRODYNAMICS

The modeling of rivers or lakes is closely related to questions of environmental protection, water supply, and ecology. One of the rudimentary requirements for numerical modeling is an understanding of the basic physical processes which must be included in the mathematical formulation. Generally, use is made of the fundamental conservation laws of mechanics and hydrodynamics. They can be formulated as integral or differential equations. Although the corresponding fundamentals are available, their application to specific situations in natural waters with their nonlinear and turbulent behavior presents some difficulties.

A discussion of the mathematical equations that exist in the literature and the existing work and successes in the application of numerical modeling techniques to solve these equations will be made first. The objective is to demonstrate the advantages and disadvantages of, and to gain some insight into, the various approaches that one may take.

The discussion is structured in the following way. It begins by summarizing those systems of fluid equations which are of broad general interest in connecting channels, rivers, and lakes. This is reviewed under the heading of mathematical equations and includes a review of the Boussinesq approximation, boundary conditions, linearization of the equations, and vertical integration.

Next, the basic numerical models available in the literature (applying these equations at various levels of complexities) are described under hydrodynamic and transport models. The review will also look at the different numerical techniques such as finite difference, finite element, method of characteristics, and cell methods. Since part of the study deals with the mixing and dispersion of pollutants in rivers, convection-diffusion river models are included. It reviews the current knowledge of side discharged pollutants into a cross flow and how they are diffused by the turbulence.

Mathematical Equations

The equations that govern the fluid behavior in rivers and lakes are reviewed first, to aid in the understanding of the discussion on the numerical models. The physical state of a river or lake can be described by the following seven macroscopic quantities which are functions of space and time:

- Velocity vector with components u, v and w   : $\vec{q}$
- Pressure   : p
- Temperature   : T
- Chemical constituent, e.g., salinity   : S
- Density   : ρ

Seven equations are required for determining these seven variables. They are obtained from the conservation laws of mechanics and thermodynamics (Krauss, 1973; Raudkivi and Callander, 1975).

I. Equation of Fluid Motion (Navier-Stokes)

For an incompressible fluid with constant viscosity,

$$\rho\left[\frac{\partial \vec{q}}{\partial t}+(\vec{q}\cdot\nabla)\vec{q}\right] = -\nabla p+\mu\nabla^2\vec{q}+\rho\vec{g}+\rho\vec{X} \tag{1}$$

where

t = time

$\vec{g} = (0,0,-g)^T$ is gravitational acceleration

$\nabla = \frac{\partial}{\partial x}\underline{i}+\frac{\partial}{\partial y}\underline{j}+\frac{\partial}{\partial z}\underline{k}$

$\nabla^2 = \frac{\partial^2}{\partial x^2}\underline{i}+\frac{\partial^2}{\partial y^2}\underline{j}+\frac{\partial^2}{\partial z^2}\underline{k}$

i, j, k = directional unit vectors

μ = the viscosity (assumed constant)

$\vec{X}$ = a vector of other body forces (Coriolis) per unit mass

If Coriolis is the only other body force in addition to gravity, then

$$\vec{X} = \omega\vec{q}$$

where

$$\omega = \begin{pmatrix} 0 & 2\Omega\sin\lambda & -2\Omega\cos\lambda \\ -2\Omega\sin\lambda & 0 & 0 \\ 2\Omega\cos\lambda & 0 & 0 \end{pmatrix}$$

Ω = the angular velocity of the earth

λ = the latitude

II. The Fluid Continuity Equation

The Fluid Continuity Equation, derived from the principle of conservation of mass, is given by (Krauss 1973, Raudkivi and Callander 1975)

$$\frac{\partial\rho}{\partial t} + (\vec{q}\cdot\nabla)\rho + \rho\nabla\cdot\vec{q} = 0 \tag{2}$$

III. The Energy Equation

The Heat Balance Equation (Krauss 1973, Raudkivi and Callander 1975) is

$$\frac{\partial T}{\partial t} + (\vec{q}\cdot\nabla)T + T\nabla\cdot\vec{q} = \beta^{T}\nabla^{2}T \tag{3}$$

where

β = (molecular) thermo-diffusion coefficient

IV. The Mass Transport Equation

The Mass Transport Equation for the chemical constituent (Krauss 1973, Raudkivi and Callander 1975) is given by

$$\frac{\partial S}{\partial t} + (\vec{q} \cdot \nabla)S + S\nabla \cdot \vec{q} = \beta^s \nabla^2 S \tag{4}$$

where

β^S = (molecular) salinity diffusion coefficient

It is assumed that the chemical constituents are nonreactive.

V. The Equation of State

The Equation of State relating fluid density to both temperature and chemical concentration (Krauss 1973, Raudkivi and Callander 1975), has the general form

$$\rho = F(\rho, T, S) \tag{5}$$

in which F is an empirical function.

VI. Reynolds Equations

The systems of Equations 1 to 5 include more types of phenomena than are often relevant to the study of natural water bodies, like rivers, lakes, and estuaries. In many natural water bodies the flow is predominantly turbulent. The Reynolds form of the governing equations is particularly valuable (Hinze 1959) for treating turbulent flow.

If q(x,y,z,t) is a physical quantity of the fluid, it can be split into a mean and a fluctuating component as given below:

$$q(x,y,z,t) = q_{mean}(x,y,z,t) + q'(x,y,z,t) \tag{6}$$

where the mean component is defined as an ensemble average while the fluctuating component is the random fluctuation of the instantaneous value about the mean value.

The application of Equation 6 to the basic Equations 1 to 5 results in the Reynolds equation in terms of the mean quantities which are of interest in this study. The resulting Reynolds equations, in tensor notation for an incompressible fluid with constant viscosity (Hinze 1959) can be written as,

Momentum:

$$\rho\left(\frac{\partial \overline{u_i}}{\partial t} + \overline{u_j}\frac{\partial \overline{u_i}}{\partial x_j}\right) = -\frac{\partial \overline{p}}{\partial x_i} + \frac{\partial}{\partial x_j}\left(\mu\frac{\partial \overline{u_i}}{\partial x_j} - \rho\overline{u_i u_j}\right) + \overline{F_i} \tag{7}$$

and,

Continuity:

$$\frac{\partial}{\partial x_i}\left(\overline{u_i}\right) = 0 \qquad (8)$$

in which

$$\overline{u}_i = \text{time mean velocity vector } (i = 1, 2, 3)$$

$$x_i, x_j = \text{spacial coordinates } (i, j = 1, 2, 3)$$

$$\rho\overline{u_i u_j} = \text{turbulent stresses, i.e., Reynolds stresses}$$

$$\overline{p} = \text{time mean static pressure}$$

$$\overline{F}_i = \text{time mean external force (i.e., gravity; Coriolis)}$$

VII. The Averaging of the Transport Equation for Turbulent Motion of a Scalar Quantity

The averaging of the Transport Equation for Turbulent Motion of a Scalar Quantity, i.e., such as temperature or salinity, for an incompressible fluid is given by Hinze (1959).

$$\frac{\partial\overline{\varsigma}}{\partial t} + \overline{u}_i\frac{\partial\overline{\varsigma}}{\partial x_i} = \frac{\partial}{\partial x_i}\left(M_D\frac{\partial\overline{\varsigma}}{\partial x_i} - \overline{u_i\vartheta}\right) + \overline{F}_\vartheta \qquad (9)$$

in which,

$$\overline{\varsigma} = \text{time mean scalar quantity per unit mass}$$

$$\overline{u_i\vartheta} = \text{turbulent correlation for the scalar quantity}$$

$$M_D = \text{molecular transport coefficient}$$

$$\overline{F}_\vartheta = \text{time mean driving force}$$

There remain, however, nonvanishing correlations of the turbulent fluctuations such as the Reynolds stresses in Equation 7 and the correlation between the turbulent fluctuations of the velocity and the scalar quantity (i.e., temperature or salinity) in Equation 9. These quantities act as additional stresses or diffusion; their elimination is the closure problem of turbulent flow.

Often, the Boussinesq approximation (Hinze 1959) is used for representing turbulence by means of eddy viscosity or eddy diffusion coefficients. The eddy coefficients can be determined by calibration against field data. In doing so, analytical approaches such as the Prandtl's mixing length hypothesis are of some help.

Recently mathematical models of turbulence have also been increasingly used (Rodi 1984). Here the eddy viscosity is related to characteristic quantities of the turbulence such as the kinetic energy, k, and the dissipation, ε. The unknowns k and ε are determined by auxiliary transport equations (Launder and Spalding 1972). These are given below in tensor notations.

$$\rho\left(\frac{\partial k}{\partial t}+\overline{u}_j\frac{\partial k}{\partial x_j}\right)=\frac{\partial}{\partial x_j}\left(\frac{v_t}{\sigma_k}\frac{\partial k}{\partial x_j}\right)+v_t\left(\frac{\partial \overline{u}_i}{\partial x_i}+\frac{\partial \overline{u}_j}{\partial x_i}\right)\frac{\partial \overline{u}_i}{\partial x_j}-\varepsilon \qquad (10)$$

and

$$\rho\left(\frac{\partial \varepsilon}{\partial t}+\overline{u}_j\frac{\partial \varepsilon}{\partial x_j}\right)=\frac{\partial}{\partial x_j}\left(\frac{v_t}{\sigma_\varepsilon}\frac{\partial \varepsilon}{\partial x_j}\right)+c_1 v_t\frac{\varepsilon}{k}\left(\frac{\partial \overline{u}_i}{\partial x_i}+\frac{\partial \overline{u}_j}{\partial x_i}\right)\frac{\partial \overline{u}_i}{\partial x_j}-c_2\frac{\varepsilon^2}{k} \qquad (11)$$

in which

$$v_t=c_\mu\frac{k^2}{\varepsilon}=\text{turbulent eddy viscosity}$$

$c_1, c_2, c_\mu, \sigma_\kappa, \sigma_\varepsilon$ are empirical constants

These empirical constants appear to change very little, even in varying applications.

Specifying Simplification Conditions

In applying the Equations 1 to 6, one or more of the following simplifications are commonly used:

1. The water is incompressible. The density is treated as constant except in the buoyancy term (the Boussinesq approximation).
2. Vertical accelerations are to be neglected.
3. The flow is quasihydrostatic, i.e., hydrostatic pressure is assumed.
4. The only external forces are gravity and Coriolis forces. The Coriolis parameter, f, is taken as constant or neglected.
5. Salinity is taken as constant or neglected.
6. Temperature is constant, i.e., stratification is neglected.

By means of the above simplifications, the following equations are obtained (Pinder and Gray 1977):

x-momentum:

$$\frac{\partial \bar{u}}{\partial t} + \bar{u}\frac{\partial \bar{u}}{\partial x} + \bar{v}\frac{\partial \bar{u}}{\partial y} + \bar{w}\frac{\partial \bar{u}}{\partial z} - f\bar{v} = -\frac{1}{\rho}\frac{\partial \bar{p}}{\partial x} + \frac{\partial}{\partial x}\left(\eta_H \frac{\partial \bar{u}}{\partial x}\right) +$$

$$\frac{\partial}{\partial y}\left(\eta_H \frac{\partial \bar{u}}{\partial y}\right) + \frac{\partial}{\partial z}\left(\eta_v \frac{\partial \bar{u}}{\partial z}\right) \quad (12)$$

y-momentum:

$$\frac{\partial \bar{v}}{\partial t} + \bar{u}\frac{\partial \bar{v}}{\partial x} + \bar{v}\frac{\partial \bar{v}}{\partial y} + \bar{w}\frac{\partial \bar{v}}{\partial z} + f\bar{u} = -\frac{1}{\rho}\frac{\partial \bar{p}}{\partial y} + \frac{\partial}{\partial x}\left(\eta_H \frac{\partial \bar{v}}{\partial x}\right) +$$

$$\frac{\partial}{\partial y}\left(\eta_H \frac{\partial \bar{v}}{\partial y}\right) + \frac{\partial}{\partial z}\left(\eta_v \frac{\partial \bar{v}}{\partial z}\right) \quad (13)$$

z-momentum:

$$\frac{\partial p}{\partial z} = -\rho_o g \quad (14)$$

Continuity:

$$\frac{\partial \bar{u}}{\partial x} + \frac{\partial \bar{v}}{\partial y} + \frac{\partial \bar{w}}{\partial z} = 0 \quad (15)$$

in which

$\bar{u}, \bar{v}, \bar{w}$ = time averaged x, y, z velocity components

ρ_o = variable fluid density

g = gravitational acceleration

f = $2\Omega\sin\lambda$ = Coriolis parameter

η_H = horizontal eddy viscosity

η_V = vertical eddy viscosity

In some large bodies of water the ratio of nonlinear inertial forces to the Coriolis forces (characterized by the Rossby number) in the momentum equations is so small that the nonlinear inertial terms can be neglected (Liggett 1975).

This assumption leads to linear momentum equations, e.g., for steady state conditions,

x-momentum:

$$-f\bar{v} = -\frac{1}{\rho}\frac{\partial \bar{p}}{\partial x} + \frac{\partial}{\partial x}\left(\eta_H \frac{\partial \bar{u}}{\partial x}\right) + \frac{\partial}{\partial y}\left(\eta_H \frac{\partial \bar{u}}{\partial y}\right) + \frac{\partial}{\partial z}\left(\eta_v \frac{\partial \bar{u}}{\partial z}\right) \tag{16}$$

y-momentum:

$$f\bar{u} = -\frac{1}{\rho}\frac{\partial \bar{p}}{\partial y} + \frac{\partial}{\partial x}\left(\eta_H \frac{\partial \bar{v}}{\partial x}\right) + \frac{\partial}{\partial y}\left(\eta_H \frac{\partial \bar{v}}{\partial y}\right) + \frac{\partial}{\partial z}\left(\eta_v \frac{\partial \bar{v}}{\partial z}\right) \tag{17}$$

The z-momentum and continuity Equations 14 and 15 are not changed.

Boundary Conditions

In order to integrate the partial differential Equations 12 to 15, boundary conditions must be prescribed at the physical boundaries of the water body. They represent the fluxes of momentum through any open boundaries (inflows and outflows from tributaries) or through the free surface (precipitation and evaporation) or through the bottom (groundwater infiltration), as well as the kinematic conditions. Normally they are chosen as follows (Liggett 1975):

Free Surface:

$$\rho\eta_v\left(\frac{\partial \bar{u}}{\partial z}\right)_{surface} = \tau_x \tag{18a}$$

$$\rho\eta_v\left(\frac{\partial \bar{v}}{\partial z}\right)_{surface} = \tau_y \tag{18b}$$

$$\frac{\partial \zeta}{\partial t} + \bar{u}_s\frac{\partial \zeta}{\partial x} + \bar{v}_s\frac{\partial \zeta}{\partial y} = \bar{w}_s \; on \; z = \zeta \tag{19}$$

$$p = 0 \quad on \; z = \zeta \tag{20}$$

in which

τ_x, τ_y = wind generated surface stresses

$\bar{u}_s, \bar{v}_s, \bar{w}_s$ = surface velocity vectors

ζ = surface elevation above the original free surface

Bottom:

$$\rho \eta_v \left(\frac{\partial \overline{u}}{\partial z} \right)_{bottom} = \tau_{bx} \tag{21a}$$

$$\rho \eta_v \left(\frac{\partial \overline{v}}{\partial z} \right)_{bottom} = \tau_{by} \tag{21b}$$

in which

τ_{bx}, τ_{by} = bottom frictional stresses

Lateral solid boundaries:

$$\overline{u} = \overline{v} = \overline{w} = 0 \tag{22}$$

Sometimes the condition of no slip is also used at the river or lake bottom.

For the surface and bottom stresses, the following empirical relationships are commonly chosen:

$$\tau_x = c_{Drag} \sqrt{\left(W_x^2 + W_y^2 \right)} W_x \tag{23a}$$

$$\tau_y = c_{Drag} \sqrt{\left(W_x^2 + W_y^2 \right)} W_y \tag{23b}$$

in which

C_{Drag} = wind drag coefficient

W_x, W_y = wind velocity vectors on the free surface

$$\tau_{bx} = C_B \sqrt{\left(\overline{u_b}^2 + \overline{v_b}^2 \right)} \overline{u_b} \tag{24a}$$

$$\tau_{by} = C_B \sqrt{\left(\overline{u_b}^2 + \overline{v_b}^2 \right)} \overline{v_b} \tag{24b}$$

in which

C_B = the bottom friction factor

$\overline{u_b}, \overline{v_b}$ = the bottom velocity vectors

The "rigid lid" is another simplifying approximation, which is often made if the free surface deviations are assumed small compared to the depth. This assumption eliminates all surface waves and replaces the boundary condition of Equation 19 with

$$\overline{w}_s = 0, on\, z = \zeta \,(\text{the free surface}) \qquad (25)$$

and the boundary condition of Equation 20 is no longer used. The "rigid lid" does not act like a solid boundary with respect to friction and shear (Liggett 1975). Surface elevations can be estimated if the pressures under the "rigid lid" are changed to equivalent water depths. The "rigid lid" approximation simplifies the computational process and allows the use of longer time steps in the unsteady free surface problems. In the discussion to follow, all the variables refer to the ensemble mean values and the bar has been dropped.

Vertical Integration

In shallow water bodies, vertically integrated equations and variables may adequately describe the situation. In this approach an estimate of the transport through any cross section can be obtained; however, the detailed information on the velocity structure is lost. Usually, the water density is assumed constant in the vertical z-direction. This and the assumption of relatively small vertical velocities and accelerations are normally implied by shallow water; consequently hydrostatic pressure is used.

Rastogi and Rodi (1978) presented the following system of depth averaged equations for modeling turbulent flow in rivers that are relatively straight:

x-momentum:

$$U\frac{\partial U}{\partial x} + V\frac{\partial U}{\partial y} = -g\frac{\partial H_T}{\partial x} + \frac{1}{\rho h}\left(\eta_H \frac{\partial \overline{\tau_{xy}}h}{\partial y}\right) - \frac{\tau_{bx}}{\rho h} \qquad (26)$$

Continuity:

$$\frac{\partial hU}{\partial x} + \frac{\partial hV}{\partial y} = 0 \qquad (27)$$

Turbulent kinetic energy:

$$\rho U_j \frac{\partial k}{\partial x_j} = \frac{\partial}{\partial x_j}\left(\frac{\nu_t}{\sigma_k}\frac{\partial k}{\partial x_j}\right) + G + P_k - \varepsilon \qquad (28)$$

$$for\, j = 1, 2$$

Dissipation:

$$\rho U_j \frac{\partial \varepsilon}{\partial x_j} = \frac{\partial}{\partial x_j}\left(\frac{\nu_t}{\sigma_\varepsilon}\frac{\partial \varepsilon}{\partial x_j}\right) + c_1 \nu_t \frac{\varepsilon}{k} G + \frac{\partial \overline{u}_j}{\partial x_i} + P_\varepsilon - c_2 \frac{\varepsilon^2}{k} \tag{29}$$

in which

$$\tau_{xy} = \rho\left(\nu_t\left(\frac{\partial U}{\partial y} + \frac{\partial V}{\partial x}\right) - z\delta_{xy} k/3\right) \tag{30}$$

$$G = \nu_t\left[2\left(\frac{\partial U}{\partial x}\right)^2 + 2\left(\frac{\partial V}{\partial y}\right)^2 + \left[\frac{\partial U}{\partial y} + \frac{\partial V}{\partial x}\right]^2\right] \tag{31}$$

$$\nu_t = c_\mu \frac{k^2}{\varepsilon} = \text{ turbulent eddy viscosity} \tag{32}$$

$$\tau_{bx} = c_f \rho U_T^2 \tag{33}$$

where

u, v, ut = depth averaged x, y, and total velocities

P_k = production of k by bed friction

P_ε = production of ε by bed friction

$c_1, c_2, \sigma_\kappa, \sigma_\varepsilon, c_\mu$ are empirical constants

The water body can be discretized into a number of vertically stratified layers, with averaged horizontal flows. As a limiting case, a single layer model is obtained when integrating over the whole depth. The use of vertically integrated quantities to predict the dispersion of a tracer in this situation is clearly less satisfactory. This is because of the highly nonuniform velocity profile. However, the total transport may still be reasonably well-predicted. In multilayered models, some improvement on this point can be expected.

For storm surges, the driving force, which is the hydrodynamic pressure, acts over the entire depth, and the vertically integrated values are expected to be representative for the local velocities, except for those close to the bottom. Finally, any inflows or outflows with a density difference are better simulated in multilayer models. It should also be pointed out that space-averaging reduces the computational effort. The dispersion coefficients used in multilayer models are normally determined empirically.

LITERATURE REVIEW OF HYDRODYNAMIC AND TRANSPORT MODELS

Except in a few restricted cases the basic mathematical equations discussed above cannot be solved analytically in a closed form. Therefore, they must be treated by approximation with the help of the discretization methods which have been developed in numerical mathematics. This leads to new problems originating from the computational treatment. Some of these can be dealt with, but a large portion has not yet been sufficiently studied, particularly in the case of nonlinear systems.

Except for simplified models (Shen et al. 1986), there are no detailed hydrodynamic or transport models currently available for connecting channels system. However, there are several well-tested modeling techniques that have been applied to the connecting waterways. These techniques can be classified as finite difference (FDM), finite element (FEM), cell methods, boundary element (BEM), and method of characteristics.

Although the governing equations for rivers and shallow lakes are similar in form, there are differences in the relative importance of the various terms in the momentum equation. For example, in fast flowing rivers the dominant forces are inertia, gravity, and bed friction, while in shallow lakes the inertia "force" is often neglected in comparison to wind, bed friction, Coriolis and gravity forces. Some portions of the Great Lakes Connecting Channels, such as the Upper St. Marys River, have characteristics of both lakes and rivers.

Finite Difference Models

The numerical models have traditionally employed the finite difference method to solve the governing differential equations. In essence, this method satisfies the governing equations by replacing derivatives by difference approximations.

Shallow Lakes or Embayments in Rivers

Some of the earlier research on the hydrodynamics of shallow surface waters are the works of Hansen (1956) and Welander (1957). Although both were considered vertically averaged equations, two different approaches have evolved from their work.

Hansen (1956) outlined the vertically averaged formulation which is still in use. His formulation includes a constant horizontal "viscosity" in the momentum equations. He used the finite difference method on a staggered grid in space and time. This scheme, with later modifications, has proven successful in many hydrodynamic problems. The staggered grid approach has the advantage that it allows the use of central differences in space and time. This feature is desirable for accuracy and numerical stability. However, one of the problems associated with the staggered grid method is the problem of representing the physical boundaries properly. It requires special treatment at the boundaries to avoid errors and instability.

Roache (1976) reviewed the finite difference method for fluid dynamics problems and emphasized the importance of satisfying the conservation laws. The requirements of stability and consistency are discussed by Roach (1976) and Richtmeyer and Morton (1967).

Leendertse (1967), using the same equation as Hansen but without eddy viscosity terms, discussed the problems encountered at the boundaries as they relate to numerical stability and accuracy. He demonstrated the importance of using centered differencing.

The treatment of nonlinear terms as usual causes severe problems. Heaps (1969) modeled the wind surges in the North Sea using staggered grid finite differencing. His numerical scheme uses the linearized and integrated dynamic equations transformed in spherical coordinates. Heaps took care to center the differences in space; he used an explicit time integration scheme. Reid and Bodine (1968) also developed a finite difference storm surge model based on the space staggered grid of Leendertse. They introduced a radiation type of boundary condition for the open ocean boundary in their work. Abbott et al. (1973) developed their models along the same line as Leendertse; however, they used a special implicit time integration method for better conservation and stability.

Simons (1972) implemented a finite difference model for the Great Lakes based on the vertically integrated equations with eddy viscosity using two space and time staggered grids simultaneously to avoid problems with the convective terms. Since Simons used a high resolution grid to discretize Lake Ontario and an explicit integration scheme, his model requires considerable computation time. Other vertically integrated models that have been applied to the Great Lakes are Rao and Murty (1970), Simons (1974, 1975, 1976), Haq et al. (1974), and Sheng (1975). These models are two-dimensional and they can only yield a first approximation to the average currents. Since the bottom stress is poorly approximated, these models do not give accurate prediction of surface elevation or the three-dimensional currents in a shallow lake such as Lake Erie where the bottom stress is relatively important (Lick 1976).

Based on the earlier work of Ekman (1905), the approach taken by Welander (1957) is noteworthy because the dependence on the vertical z-coordinate is determined analytically. This is a steady-state model that is used to predict wind induced circulation in shallow lakes (Liggett 1969). The "rigid lid" assumption is used whereby the water surface is assumed fixed. The other assumptions in the model are no horizontal momentum diffusion, negligible nonlinear terms, and negligible density stratification; vertical eddy viscosity is constant; and all velocities vanish on the bottom (no-slip). The solution proceeds by expressing the vertical z-dependence as a Fourier series. Next, the vertically integrated equations are solved by introducing a stream function that satisfies the continuity equation. The pressure is eliminated from the momentum equations, leaving one partial differential equation in the stream function. This equation with the proper boundary conditions is solved using finite differencing.

Unsteady "rigid lid" models have been used in modeling currents in large lakes by Liggett (1969), Paul and Lick (1974), Bennett (1977), and Schwab et al. (1981). Since the vertical velocity is assumed zero at the undisturbed location of the free surface, it eliminates the motion and time scales associated with the surface gravity waves and allows a much larger numerical time step and reduced computation times.

Free surface models are computationally inefficient since the maximum time step is limited by the time it takes the surface gravity wave to travel between two adjacent horizontal grid points. Several free surface models have been applied to the Great Lakes (Freeman et al. 1972, Kizlauskaus and Katz 1974, Haq et al. 1974, Sheng 1975, Schwab et al. 1981).

To reduce the computational time, two-mode free surface models have been developed (Simons 1972, 1974, 1975; Sheng and Lick, 1977). In these two-mode models, the free surface elevation is treated separately from the internal, three-dimensional flow variables. The free surface elevation and vertically integrated quantities are calculated using vertically integrated equations of motion; the limiting time step for this part of the calculation is that associated with the surface gravity waves. The internal variables are calculated in such a manner that the effect of surface gravity waves is avoided and surface gravity waves no longer limit this part of the calculation. Sheng, Lick, Gedney, and Molls (1978) discuss the applications of the "rigid lid" and the free surface models to the Great Lakes using field data for comparison. Long-term time-averaged circulation computed by the two models agree well in periods of strong wind, but differ appreciably in periods of light wind and active seiching.

Convection-Diffusion River Models

The discharge of industrial and municipal outfalls are mostly of the shore-based type. The discharge jet is normally bent by the momentum of the river flow. Near the bank, a recirculation zone can form as the river water is entrained into the expanding jet. In this zone, a strong three-dimensional flow field is apparent. Beyond this relatively small recirculation zone, the flow field follows the dominant flow direction of the river. This observation is of great significance to the modeler because of the complexities of a three-dimensional recirculating flow field. Rodi et al. (1981) categorize the flow regions as near-field and far-field regions.

The near-field region is the region that extends a relatively short distance downstream from the outfall wherein the influence of the jet discharge still affects the flow field. In this region, the generation of turbulence is partly due to the shear layers induced by the jet in the cross-flowing stream and due partly to the river bed. Usually the stratification and vertical nonuniformities in this region are eroded by vertical mixing. The governing equations in this region are three dimensional and elliptic in nature. The three dimensional flow field and the elliptic equations require an iterative solution procedure leading to large computing times. To overcome these difficulties, the near-field region is often simulated by integral methods which require profile assumptions and entrainment parameters.

Shirazi and Davis (1974) and Abdel-Gawad (1985) had used a three-dimensional integral model referred to as the PDS model. The model requires specification of shape functions for the lateral and vertical temperature and velocity profiles. Selection of these shape functions enables one to integrate the basic conservation equations over the cross section of the jet. Abdel-Gawad (1985) also used another near field model referred to as the MIT model. The model was developed by Stolzenbach and Harleman (1971), and like the PDS model is based on the integral approach. In contrast to the PDS model that uses the Gaussian distribution, the MIT model is developed using a "top hat" distribution which is defined in terms of "jet averaged" properties.

Demuren and Rodi (1983) succeeded in modeling the near-field region with a three-dimensional model using the "κ-ε" turbulence model to determine the turbulent stresses. The model employs differential transport equations (κ and ε equations) for the turbulent momentum transport quantity, like the turbulent stresses. It traces the transport, diffusion, generation, and decay of the turbulent energy level "κ" and the transport, diffusion, generation, and decay of a characteristic length scale, represented by the rate of turbulent energy loss to heat energy, "ε". The model performed well for isothermal discharges; however, they did not include the buoyancy terms in their model.

The far-field region begins where the near-field region ends and extends as far downstream as the effluent concentration is detectable. This is a vertically well-mixed region and the turbulence is predominantly generated from the river bed. Under such well-mixed situations when temperature and pollutant concentration vary little over the river depth, it is accurate for practical purposes to use two-dimensional depth-averaged equations (Rodi et al. 1981). Except for small recirculation regions, the flows in the connecting channels are predominantly in the main flow direction and are strong enough to prevent upstream diffusion of contaminants. Thus the steady-state mass transport is controlled by downstream convection and lateral turbulent mixing (Rodi et al. 1981).

When downstream influences can be predetermined, e.g., by using field observations of water levels or by applying backwater programs like HEC-2 (U.S. Army Corps of Engineers — Water Surface Profile model), the transport equations can be made parabolic in the main-flow direction, i.e., the solution at a certain cross section does not depend on the solution at cross sections located downstream nor on the conditions at the outflow boundary. Therefore, in a numerical scheme, the equations can be integrated by marching from one cross section to the next, starting with given initial conditions at the cross section furthest upstream. The forward-marching procedure is extremely economical since all variables have to be stored only at grid nodes in one cross section.

Knowledge of the transverse mixing coefficient is still very limited. Some reseachers have suggested that it is a function of the bulk flow and the channel geometry (Yotsukura and Sayre 1976, Lau and Krishnappan 1981). Most of the advective-diffusion equations describing the depth-averaged concentration of a pollutant require field data to estimate the tranverse mixing coefficient. This coefficient accounts for the turbulent diffusive transport as well as the transport caused by differential advection. Fisher (1969) derived a relationship for the transverse dispersion coefficient for rivers. Data from the Missouri River based on the work of Yotsukura et al. (1970), together with data from flume experiments by Fisher (1969), have shown that Fisher's equation predicts the correct trend but is not sufficiently close as yet for unrestricted application in the field [Yotsukura and Sayre (1976)].

Studies in meandering channels have shown that the transverse mixing coefficient tends to vary periodically in the longitudinal direction and is higher than in straight channels.

Due to the inherent problems involved in trying to simulate the process of pollutant diffusion, no single model is available to simulate the entire pattern of contaminant spread. In the case of natural streams, numerous models for predicting the dispersion of the pollutant are available from the works of Fisher (1967, 1968, 1969, 1973), Yotsukura and Sayre (1976), Lau and Krishnappan (1981), and McCorquodale et al. (1983).

Finite Element Models

The finite element method has emerged as a relatively powerful method in recent years. In this method, the function satisfying the governing equations and boundary conditions is approximated by piecewise polynomials. Very flexible grid discretization is by virtue of this method. Zienkiewicz (1977) has collected a number of such applications along with some more recent fluid flow problems. A survey of the finite element method in continuum mechanics, with a discussion of the Galerkin expression for a Newtonian fluid, is also given by Connor (1973).

Shallow Lakes

Shallow water circulation has attracted quite a few finite element modelers. Gallagher et al. (1973) analyzed steady wind-driven circulation for shallow lakes using the rigid lid equations. Taylor and Davis (1972) solved the vertically averaged equations for constant density. They used cubic isoparametric elements to discretize their study area. For the time integrations, they compared the effectiveness of using a fourth order predictor-corrector method, the trapezoidal rule, and finite elements in time. Grotkop (1973) treated the same problem using linear finite elements in space and time; the trapezoidal rule was computationally more efficient for their case but the other methods were more accurate.

A review of finite element models for fluid flow was prepared by Norton et al. (1973). Pinder and Gray (1977) have documented finite element techniques for modeling lakes and estuaries. Connor and Brebbia (1976) also authored a book that discusses finite element techniques in fluid flow. It includes chapters on the simulation of lake hydrodynamics and transport of pollutants using finite element methods. Lynch and Gray (1979) and Kinmark and Gray (1984) had devoted their efforts to developing finite element techniques using nine node elements to solve the wave equation for estuaries and lakes with tidal boundary conditions. Their work was directed towards improving stability and controlling numerical dispersion in their solutions. Others who have applied finite element methods to lake circulation include Cheng (1970, 1972), and Chih-lan et al. (1976).

Most of the recent researchers have concentrated on refining their finite element techniques to speed up computation time, improving accuracy or reducing numerical dispersion. The mathematical formulations are basically similar to earlier works. However, Laible (1984) attempted to resolve the vertical dimension by the use of higher order shape functions to approximate the natural current profile; this approach was attempted previously by Ebeling (1977) using cubic elements. Recently, Ibrahim and McCorquodale (1985) applied the finite element method to Lake St. Clair with some success. They introduced optimum upwinding finite element schemes to handle the convective terms in their transport model and also developed a partial slip boundary condition for their hydrodynamic model.

Rivers

Finite Element Methods have been used for both one-dimensional and two-dimensional steady and unsteady flow in rivers. Moin (1988) surveyed the one-dimensional FEM models for unsteady flow. Moin (1988) has also developed and tested one-dimensional FEM models for unsteady flow in rivers. Froehlich (1988) developed a two-dimensional FEM models for rivers of variable depth and cross section; this model treats steady flow and uses the "κ-ε" turbulence model.

The problem of numerical stability arises in both one- and two-dimensional FEM solutions; this is due in part to the large convective term relative to the "diffusion" term in the equation of motion. Researchers have suggested a number of methods of overcoming this problem, for example: Moin (1988) used a moving finite element method for the dam break problem; Katopodes (1984) used a dissipative Galerkin scheme which was successfully applied to the Detroit River (UGLCC 1988); Huyakorn (1977) is one of several researchers to have used upwinding techniques to obtain stable solutions to the transient transport equation.

Cell Models

Cell models divide the water body into finite volume segments (cells) that are considered to be well-mixed. Most cell models require a separate hydrodynamic model to assign the intercell advective flows. The problem of different time scales and grid compatibility occurs when transferring information from the hydrodynamic model to the cell model. Diffusive mass fluxes can be estimated from turbulence modeling or directly from field data. Ideally, intercell fluxes are initially based on advection and assumed diffusion coefficients and then these coefficients are adjusted until there is conformity between observed and calculated distributions of a conservative substance such as chloride. A system of ordinary differential equations is derived by applying the conservation of mass equation to each cell to express the time rate of change of concentration of the contaminant. This method can be applied to one-, two-, or three-dimensional flows; however, the specification of the input data and the computation time may limit the number of cells and/or the complexity of their arrangement. In some cases the thermal energy equation must be solved along with the hydrodynamic equations. Thomann et al. (1975) applied this method to Lake Ontario.

Several cell or compartmental models (Di Toro et al. 1980, 1982) have been developed for the simulation of pollutant transport in rivers, estuaries, and lakes. Many of these models are used for aquatic ecological modeling, which is intrinsically related with water quality modeling. Lorenzen et al. (1974) reviewed various such models which were used to evaluate the Great Lakes. One class of models is represented by the phytoplankton model developed by Di Toro et al. (1975) for western Lake Erie. Richardson and Bierman (1976) applied similar models to Saginaw Bay. Others who have applied these type of models include Canale et al. (1974) to Grand Traverse Bay and Scavia et al. (1976) to Lake Ontario.

Richardson et al. (1977) and Ambrose et al. (1988) of the U.S. EPA (Environmental Protection Agency), as well as researchers at Manhattan College, New York, have developed and documented a very flexible cell model commonly referred to as WASP4 (Water Analysis Simulation Program) for rivers, estuaries, and large lakes. The WASP4 package includes three components, namely DYNHYD4, EUTRO4, and TOXI4. The hydrodynamics can be either specified by the user or computed from DYNHYD4 which is a one-dimensional link-node hydrodynamic model. Depending on the problem being studied, the user can use either EUTRO4 and TOXI4 as the fate and transport submodels to link to the main WASP4 model. This will complete the "full" model for simulating the movement and interaction of the pollutant within the water body. EUTRO4 supplies the chemical kinetics for conventional pollutants involving dissolved oxygen, biochemical oxygen demand, and nutrients related to eutrophication. TOXI4 supplies the chemical kinetics for toxic pollutants involving organic chemicals, metals, and sediment. After calculation of toxicant concentrations in space and time for the water body, a Food Chain Model (e.g., Connolly and Thomann 1985) can be used to predict uptake and accumulation in biota. The Food Chain Model can be used as a part of the WASP4 system (Ambrose et al. 1988).

Method of Characteristics

An outline of the method of characteristics for the one-dimensional shallow water equations is given by Liggett and Woolhiser (1967). The advantage of this approach is that the original system of partial differential equations can be written as ordinary differential equations on the characteristics. However, these characteristics are, in general, curved and time dependent, thus making a solution more difficult to interpret. Although the same methodology can be extended to two-dimensional flow there seems to be no incentive for such work as the curvature of the characteristics makes it a difficult bookkeeping and interpolation process to obtain a useful solution. Compared with well-established finite difference methods, there does not seem to be a great advantage in pursuing the method of characteristics for two-dimensional flow.

Boundary Element Models

The boundary element or boundary integral method has been used in structural mechanics for some time now. Brebbia et al. (1980) used it to solve a simplified lake circulation problem. The method can be used to reduce the input data, the number of unknowns, and consequently the computer storage requirements. However, its superiority at the moment is limited to potential flow problems, such as Darcy flow in porous media and irrotational flow or flow described by the Poisson equation. The method loses its advantage if depth variation in the lake is significant or if the nonlinear terms are included in the governing equations.

PHYSICAL-CHEMICAL CONSIDERATIONS

Since the ultimate objective in modeling the hydrodynamics of the connecting channels is to simulate the movement and fate of toxic substances, a short review of the physical-chemical processes of such substances is appropriate. To better understand the terms used in this study, a knowledge of the types and sources of pollutants is necessary. In their report to the International Joint Commission (IJC), the Great Lakes Water Quality Board (1982), has defined guidelines for pollutants under four broad categories:

1. "Conventional" pollutants, e.g., BOD, COD, nutrients, oil and grease, and bacteria
2. Metals, e.g., mercury, lead, zinc, iron, and cadmium
3. "Conventional" toxic substances, including phenol, cyanide, ammonia, and chlorine
4. Persistent organic toxic substances, complex organic chemicals, which persist and can bioaccumulate

The sources of pollutants can be identified in several general categories: municipal and industrial discharges, waste disposal sites, combined sewer overflows, urban land runoff, agricultural land runoff, atmospheric, and in-place pollutants.

There are three important features that separate toxic substances from the more conventional pollutants. Thomann and Mueller (1987) listed these features as:

1. Certain toxicants tend to sorb to particulates in the water column and bed sediments.
2. Certain toxicants can accumulate in biota.
3. Certain toxicants tend to be toxic at relatively low water concentrations in the µg/L or ng/L level.

Some physical-chemical phenomena that affect toxicants in the water column and the sediment are as follows:

1. Sorption and desorption of the toxicant between dissolved and particulate forms in the water column and sediment.
2. Settling and resuspension mechanisms of sediment-sorbed toxicants between the sediment and the water column.
3. Vertical diffusive exchange between the sediment and the water column, such as pore water diffusion and percolation. Also pore water advection due to pore water flows into and out of the sediment bed.
4. Net loss of the toxicant due to biodegradation, volatilization, photolysis, and other chemical and biochemical reactions.
5. Net gain of the toxicant due to chemical and biochemical reactions.
6. Water column transport of the toxicant due to advective flow transport and dispersive mixing and lateral sediment transport in the form of bedload transport of sediment-sorbed toxicants.
7. Net deposition and loss of the toxicant to deep sediments due to sedimentation and compaction over the years.

It is important to have an idea of the time to equilibrium of the transformation mechanisms when developing the hydrodynamic and transport models so that appropriate spatial and temporal scales can be set. These transformation processes are usually grouped under fast and slow reactions. Fast reactions have characteristic reaction times on the same order as the model time step and are sometimes treated with the assumption of local equilibrium. Slow reactions have characteristic reaction times much longer than the model time step. These are usually handled with the assumption of local first order kinetics using a lumped rate constant that is based on the summation of several process rates.

This brief review indicates the type of the complex and dynamic chemical-sediment interactions that take place in a water body such as a river or a lake. A full treatment of the kinetic processes of sorption-desorption, biodegradation, photolysis, hydrolysis, volatilization, and oxidation is beyond this chapter. These physical-chemical processes are discussed in other references such as Mills et al. (1982, 1985), Thomann and Mueller (1987), and Ambrose et al. (1988).

STAGES IN MODEL DEVELOPMENT

A preliminary study is made of the physical geometry of the study area to be modeled to set the modeling boundaries. For instance, control structures are chosen to avoid the problem of dynamic hydraulic boundary conditions. Other model boundaries alternatives include extending the boundaries to or beyond the limits of the water quality impact zones.

The resolution of the spatial and temporal scales for the models is decided by the nature of the problem to be analyzed. A compromise between the accuracy of important simulation variables, spatial resolution, numerical stability, and reasonable computational time must be resolved.

Next the model grid is constructed for the study area. It is necessary to identify sampling stations, points of interest, and waste load sources so that the best alignment of the grid network is used. If results generated by the hydrodynamic model are to be stored for use by the water quality model, then grids for both the models and their time steps must be compatible though not necessarily identical.

The model development proceeds through four general stages involving hydrodynamics, mass transport, contaminant transformations, and the aquatic food chain. A hydrodynamic model is needed to calculate the water circulation and transport characteristics of the connecting channels, rivers, and lakes. This stage essentially answers the question of where the water in the channel, river, or lake goes. The second stage addresses the question of where the materials (i.e., the pollutants) in the water are routed. The third stage, which is linked to the second, answers the question of how the material in the water and sediment is transformed. The food chain model helps to answer questions about the possible impacts on the ecosystem.

Each stage of the model development requires calibration, and verification. Model calibration is the "tuning" of the model to produce output that agrees with the observed data. This may involve a combination of laboratory studies and field monitoring. For the modeler, it involves parameter estimation and possibly reformulation of the model. Model verification is testing of the calibrated model using additional field data, i.e., a set of field data not used in the original model calibration, preferably under different external conditions to further examine model validity.

APPLICATION — MODELING OF CONTAMINANTS IN THE ST. MARYS RIVER

St. Marys River is a vital navigational channel link between Lake Superior and Lakes Huron and Michigan via Lake Nicolet. It has been indentified as an area of concern by the International Joint Commission (IJC). Figure 1 shows the municipal intakes, outfalls and industrial waste discharges for the river system. The freshwater of the St. Marys River provides drinking water, fisheries habitat, hydroelectric power, and water for the local industries. Improving the quality of these waters is important to the well-being of the communities on both the Canadian and the U.S. sides of the St. Marys River.

The Ontario Ministry of Environment (MOE) through its MISA program conducted extensive field studies on the St. Marys River. These studies will result in regulation of major industrial and municipal dischargers to the Ontario side of the waterway. There is a need for efficient and workable environmental management models to assist the MOE in setting water quality-based load allocations and to help the local industrial and municipal dischargers in their self-monitoring and control efforts.

The modeling of the Lower St. Marys River, shown in Figure 1, will be discussed in detail. The river is discretized into a system of reaches in order to handle islands, confluences, diversions, and the river curvature. A schematic of the river system is shown in Figure 2. The corresponding physical discretization of the river into reaches is shown in Figure 3. It should be pointed out that each reach is further subdivided at the user's discretion, into additional cross sections in the longitudinal direction with each section consisting of 15 lateral grid points.

Model Description

KETOX has evolved from the earlier κ-ε model (McCorquodale et al., 1986) which was applied to the St. Clair, Detroit, and Niagara Rivers. KETOX is a model with several submodels. The coupling of the submodels is controlled by a directory file which describes the river system. It has a depth-averaged hydrodynamic submodel which is coupled to a convection-diffusion (mixing) submodel that simulates river dispersion of contaminants from single or multiple discharges. There is an option to call a turbulence transport submodel that calculates the variation of the lateral dispersion coefficients at grid locations across each cross section of the river using the turbulence transport equations κ and ε (Rastogi and Rodi, 1978). There are submodels to "split" the flow at diversions or islands and to "combine" the flow at confluences. Lastly, it has a toxic substances submodel to simulate the chemical-sediment interaction and transformation of toxic pollutants (Di Toro et al. 1982, 1984). This submodel component of KETOX will not be presented at this time.

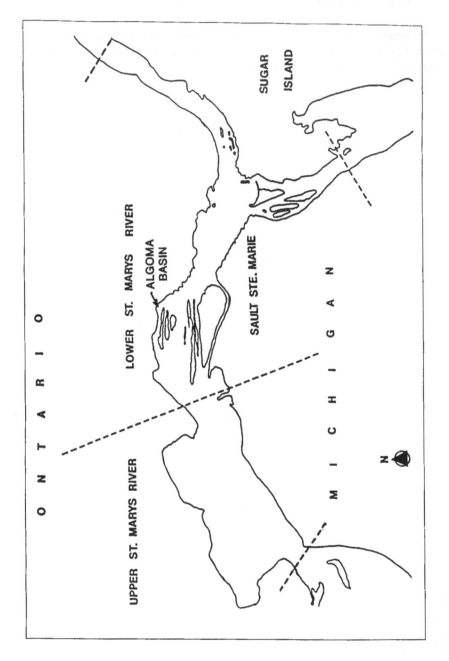

Figure 1.   St. Marys River - Algoma Terminal Basin.

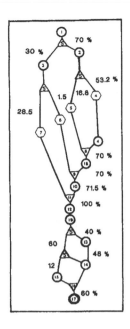

Figure 2.   System diagram for the Lower St. Marys River.

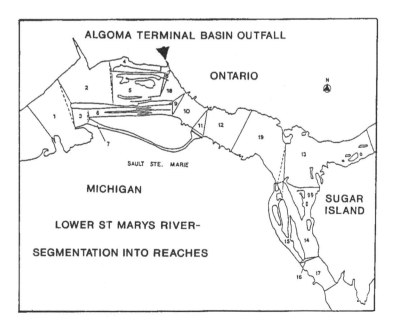

Figure 3.   Definition of reaches in the Lower St. Marys River for the
KETOX model.

Hydrodynamic Submodel

A simplified hydrodynamic submodel was used (McCorquodale et al. 1983) for KETOX. The model assumes that the lateral depth profiles and river flow rates are available and can be used to obtain average velocity in the longitudinal (flow) direction. The model accounts for river curvature effects using an analytical approach (Chang 1983). The Manning's equation is used to account for the vertical momentum transfer. To account for the side effects of the river banks, a shape function is incorporated into the model. The vertically-averaged longitudinal velocity is then given by

$$U(y) = \chi(y)\frac{h^{2/3}S_o^{1/2}}{n}\sqrt{\frac{r_c}{r_{loc}}}\qquad(34)$$

in which

$\chi(y)$ = a shape factor to account for the river bank effects
h = an average depth at a specified grid point on the river cross-section
S_o = an effective frictional slope
n = Manning's roughness factor
r_c = radius of curvature of the center line of the river width
r_{loc} = r_c = (W/2 + y) = a local radius of curvature
y = lateral distance of the local grid point from the source
W = width of the river

The boundary conditions are at $y = 0$, $U(y=0) = 0$ and at $y = W$, $U(y=W) = 0$. The relationship for the shape factor is given by

$$\chi(y) = \left[-\left(\frac{y}{W} - \frac{1}{2}\right)^2 + \frac{1}{4} \right]^{n_1}\qquad(35)$$

in which n_1 = an empirical exponent to be calibrated. The values for n_1 and n are adjusted to best fit the field velocity data [U.S. Army Corps of Engineer (COE 1984)]. In the momentum option, the longitudinal momentum equation is solved in the stream function coordinate (Rodi 1984). This option incorporates the dissipation and dispersion of the excess momentum from upstream discharges (Rodi 1984).

The continuity equation to be satisfied during this adjustment is given by

$$Q = \int_0^W U(y)h.\,dy\qquad(36)$$

Equation 36 is integrated numerically to ensure that mass is conserved from section to section along the river and to transform the output from stream function coordinates to physical locations in the river.

Pollutant Dispersion SubModel

The dispersion submodel is based on the work of Lau and Krishnappan (1981). This model solves the transverse mixing of a pollutant discharged into a river. The stream function form of the dispersion submodel is given by

$$\frac{\partial C}{\partial x} = \frac{1}{Q^2} \frac{\partial}{\partial \omega} \left[\left(U h^2 m_x E_y \right) \frac{\partial C}{\partial \omega} \right]$$ (37)

in which

C = pollutant concentration
x = longitudinal distance coordinate following the meander of the river
Q = river discharge
ω = dimensionless stream function value
m_x = a coefficient for the coordinate system
E_y = turbulent mixing coefficient in the y-direction (normal to streamlines)

The boundary conditions can be simply stated as $\omega = 0$ on the left bank and $\omega = 1$ at the right bank of a river cross section.

Model Testing and Results

Field measurements for phenol concentration levels in the river were conducted in 1974 and 1983 (Hamdy et al. 1978, Ontario MOE 1986) and river velocities in 1984 (Ontario MOE 1986).

Model Calibration and Verification

The hydrodynamic component of KETOX was calibrated using U.S. Army Corps of Engineer (COE, 1984) field data based on current measurements and drogue surveys. Both the default velocity distributions and the momentum corrected velocities were computed for the river. The momentum equation [Equation (26)] was used for the final calibration because of zones of jet-like flow from the hydroelectric stations. The default velocity profile, the momentum corrected profile, and the measured velocities are shown in Figure 4. Since all current measurements and drogue survey data were used for calibrating the model, the hydrodynamic model cannot be considered to be verified at the moment.

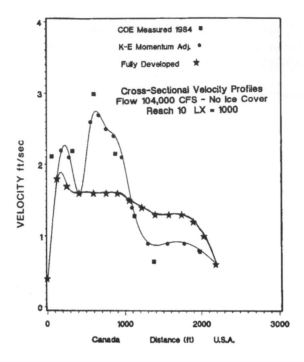

Figure 4.   Comparison of the measured and modeled velocities in reach 10 of the Lower River.

The pollutant dispersion submodel of KETOX was calibrated using the 1974 Ontario MOE (Hamdy et al. 1978) data on phenol loading and measurements along the river including transects. In order to represent the variation in loadings, multiple computer runs were made to establish the upper and lower bounds on the predictions. A comparison of the 1974 measured values and the predicted concentrations along the Canadian shoreline starting from the Algoma Terminal Basin outfall is shown in Figure 5. Most of the predicted values fall within the 95% confidence band indicating a good calibration of the submodel. The same plot shows field measurements from from 1983 (Ontario MOE 1986). The 1983 data and model prediction, Figure 6(b), indicated that KETOX and TOXI4 performed well.

Discussion

The predictive capability of KETOX provides a means for the environmental managers to screen management alternatives. For instance, the model can be used to predict the relative differences in water quality to be expected from wastewater management changes such as consolidating individual wastewater treatment plants into regional facility, relocating existing wastewater dis-

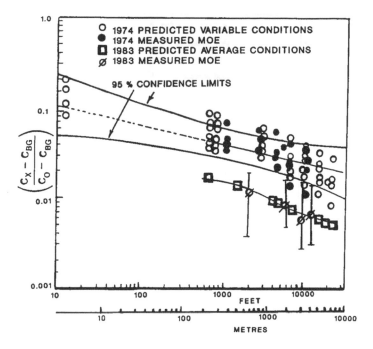

Figure 5. KETOX predicted and Ontario MOE measured concentrations of phenol along the Canadian shoreline of the Lower St. Marys River for 1974 and 1983.

charges, or incrementally reducing wastewater loads. The effects of nonpoint source load reductions can also be evaluated.

Application of the Model to Load Allocation

A mixing zone (MZ) is defined by the Ontario Ministry of Environment (1984) as "an area of water contiguous to a point source where the water quality does not comply with the Provincial Water Quality Objectives (PWQO)". This zone should be minimized in order to protect aquatic life as well as other users. Limitations on mixing zones are established on a "case by case" basis with the size of a MZ limited by the nearest downstream user. The MZ should not be a "barrier to the migration of fish and aquatic life".

The MZ is used in the allocation of loads for a particular outfall. The load is set so that the applicable guideline is satisfied at an acceptable compliance level at the edge of the MZ. Outside of this MZ there should be no long-term (chronic) effects on the aquatic life, within the MZ, the conditions should not be rapidly lethal to aquatic life. This is accomplished by using the model to establish a system response curve for the edge of the MZ, i.e., a concentration versus load

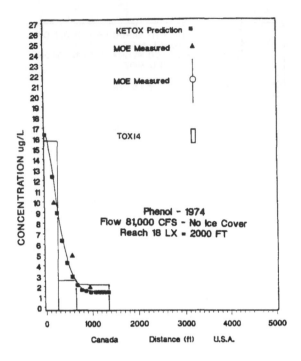

Figure 6(a).   Comparison of phenol concentrations predicted by KETOX
and TOXI4 with concentrations measured by the Ontario
Ministry of the Environment in 1974.

curve as shown Figure 7. Figure 7 shows the most probable relationship between
load and concentration at a distance of 300 m (984 feet) from the outfall for a
contaminant Y with PWQO of 1 µg/L. Entering this curve with the PWQO yields
the median load that would satisfy the objective 50% of the time (50%
compliance). The median monthly load for 50% compliance in this example is
22 kg/day. The statistical nature of the controlling variables (river flow, contaminant load, and calibration errors) of the model can be used to estimate the load
that would satisfy the objective for a specified percent of time. Figure 7 also
shows the 95% compliance response curve which yields a 95% compliance
median load of 10.6 kg/day.

Application to Impact Zones

KETOX can be used to investigate Impact Zones for pollutants under
different loading scenarios. For the Lower St. Marys River, the Impact Zone
investigation was conducted to assess the problem of transboundary pollution

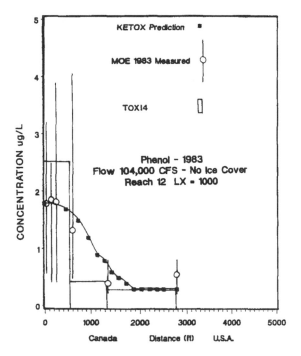

Figure 6(b).   Comparison of phenol concentrations predicted by KETOX and TOXI4 with concentrations measured by the Ontario Ministry of the Environment in 1973.

between Canada and the U.S. A volatile chemical, referred to as Y is used to show the application of the model to an Impact Zone study for different loading scenarios at the Algoma diffuser. The PWQO for chemical Y is 1 µg/L. For steady-state loading, isoconcentration maps can be developed with longitudinal resolution of the order of 50 ft (15 m) and lateral resolution as low as 1% of the flow in the reach. This permits a reasonably accurate Impact Zone to be defined so that various loading scenarios can be compared and evaluated.

Figure 8 is a plot showing the results for a typical summer impact when the initial loads the Algoma diffuser are 20 kg/day and 100 kg/day. The Algoma diffuser discharge was 4.2 m³/s with an average summer flow of 2,464 m³/s. The Impact Zone for the 100 kg/day extends from the Algoma terminal basin to the easterly Sewage Treatment Plant (STP) in the North Channel. The extent of the Impact Zone is shortened to the near-field zone of the diffuser when the load is reduced to 20 kg/day. The 1 µg/L isoconcentration line does not cross the International Boundary for either load.

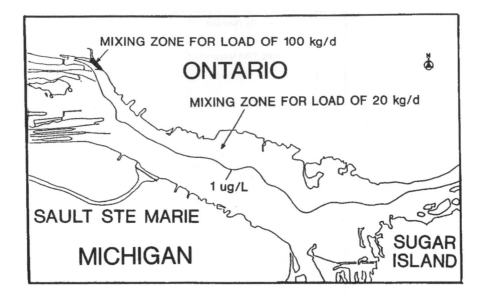

Figure 7.   Load response curve for chemical Y at the boundary of a mixing zone.

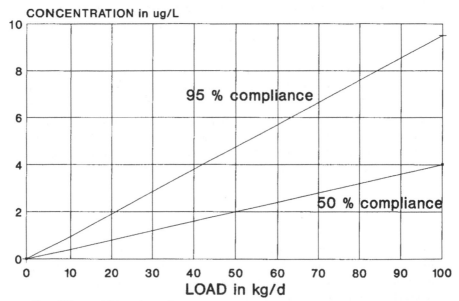

Summer Conditions- 300 m downstream of Outfall

Figure 8.   Impact zones for chemical for the Lower St. Marys River.

REFERENCES

Abbott, M.B., Damsgaard, A. and Rodenhuis, G.S. 1973. "System 21, Jupiter," *J. Hydraulic Res.*, 11 (1) (1973).

Abdel-Gawad, S.T. "Mixing and Decay of Pollutants from Shore-Based Outfalls Discharging into Cross-Flowing Streams," Ph.D. thesis, Windsor University, Windsor, Ontario (1985).

Ambrose, R.B., T.A. Wool, J.P. Connolly and S.W. Schanz. "WASP4, A Hydrodynamic and Water Quality Model – Model Theory, User's Manual and Programmer's Guide," U.S. CEAM, EPA Report-600/3-87/039 (January 1988).

Aris, R. "On the Dispersion of a Solute in a Fluid Flowing Through a Tube," *Proc. Royal Society*, A, 235:67-77 (1956).

Bennett, J.R. "A Three Dimensional Model of Lake Ontario's Summer Circulation, I: Comparison with Observations," *J. Phys. Oceanogr.* 7:591-601 (1977).

Brebbia, C.A., and L.C. Wrobel. *Steady and Unsteady Potential Problems using the Boundary Integral Method. Recent Advances in Numerical Methods in Fluids, Vol. 1* (Swanson, U.K.: Pineridge Press Ltd., 1980).

Canale, R.P., D.F. Hineman and S. Nachiappan. *A Biological Production Model for Grand Traverse Bay. Technical Report No. 37*, (Ann Arbor, MI: Sea Grant Program, University of Michigan, February 1974).

Chang, H. "Expenditure in Curved Open Channels," *J. Hydraul. Div.*, ASCE. 109 (7):1012-1022 (1983).

Cheng, R. T., and C. Tung. "Wind Driven Lake Circulation by the Finite Element Method," Proc. 13th Conf. Great Lakes Research Int. Assoc. Great Lakes Res. (1970).

Cheng, R. T. "Numerical Investigation of Lake Circulation around Islands by the Finite Element Method," *Int. J. Numer. Methods Eng.* 5:(1) (1972).

Chih-Ian, S., J.H. Pohl and G. Shih. "Wind Driven Circulations in Lake Okeechobee," 1st International Conference on Finite Element in Water Resources, Princeton University (London: Pentach Press, 1976).

Connolly, J. P., and R.V. Thomann. "WASTOX, A Framework for Modeling the Fate of Toxic Chemicals in Aquatic Environments. Part 2: Food Chain." U.S. EPA, Gulf Breeze, FL and Duluth, MN. (1985).

Connor, J.J. *Fundamentals of Finite Element Techniques*, (London: Butterworth & Co. Ltd., 1973).

Connor, J.J., and Brebbia, C.A., *Finite Element Techniques for Fluid Flow* (London: Butterworth and Co. Ltd., 1976).

Delos, C.G., W.L. Richardson, J.V. DePinto, R.B. Ambrose, P.W. Rodgers, K. Rygwelski, J.P. St. John, W.L. Shaughnessy, T.A. Faha and W.N. Christie. "Technical Guidance Manual for Performing Waste Load Allocations, Book II. Streams and Rivers, Chapter 3, Toxic Substances," U.S. EPA Report-440/4-84-022 (1984).

Demuren, A.O., and W. Rodi. Side Discharges into Open Channels: Mathematical Model. *J. Hydraul. Div.*, ASCE, 109 (12):1707-1722 (1983).

Di Toro, D.M., D.J. O'Connor, and R.V. Thomann. "Phytoplankton - Zooplankton - Nutrient Interaction Model for Western Lake Erie," in *Systems Analysis and Simulation in Ecology, Vol. III*, B.C. Patten, Ed. (New York: Academic Press, Inc., 1975), pp. 424-473.

Di Toro, D.M., and W. Matystik. "Mathematical Models of Water Quality in Large Lakes. Part 1: Lake Huron and Sainaw Bay," U.S. EPA Report-600/3-80-056 (1980).

Di Toro, D.M., D.J. O'Connor, R.V. Thomann, and J. St. John. *Simplified Model of the Fate of Partitioning Chemicals in the Aquatic Environment* (Ann Arbor, MI: Ann Arbor Science, 1982).

Di Toro, D.M., J. Fitzpatrick and R.V. Thomann. *Water Quality Analysis Simulation Program (WASP) and Model Verification Program (MVP) — Documentation* (Westwood, NY: Hydroscience Inc., 1983).

Di Toro, D.M., and P.R. Paquin. "Time Variable Model of the Fate of DDE and Lindane in a Quarry," *Environ. Toxicol. Chem.*, 3:335-353 (1984).

Ebeling, H. "Berechnung der Vertikalstruktur wind und Gezeitenerzeugter Stromungen nach der Methode der finiten Elemente," *Fortschritt-Berichte FDI-Zeitschrn.* 4 (32) (1977).

Ekman, V.W. "On the Influence of the Earth's Rotation on Ocean Currents." *Arkiv. Met. Astron. Fysik.* 2 (4) (11):52(1905).

Finlayson, B. *The Method of Weighted Residuals anad Variational Principles* (New York: Academic Press, 1972).

Fisher, H.B. "The mechanics of Dispersion in Natural Streams," *J. Hydraul. Div., ASCE* HY6:187-216 (1967).

Fisher. H.B. "Dispersion Predictions in Natural Streams," *J. Sanitation Eng. Div., ASCE* 94:927-943 (1968).

Fisher, H.B. "The Effect of Bends on Dispersion in Streams," *Water Resour. Res.* 5 (2):496-506 (1969).

Fisher, H.B. "Longitudinal Dispersion and Transverse Mixing in Open Channel Flow," *Ann. Rev. Fluid Mechanics.* 5:59-78 (1973).

Freeman, N.J., A.M. Hale and M.B. Danard. "A Modified Sigma Equations Approach to the Numerical Modeling of the Great Lakes Hydrodynamics," *J. Geophys. Res.* 77:1050-1060 (1972).

Froehlich, D.C. "Finite Element Surface-Water Element Surface-Water Modeling System: Two-Dimensional Flow in a Horizontal Plane — Users Manual," Pub. No. FHWA-RD-88-177, Federal Highway Administration.

Gallagher, R.H., J.A. Liggett and S.K.T. Chan. "Finite Element Shallow Lake Circulation," *J. Hydraul. Div., ASCE,* 99:1083-1096 (1973).

Gedney, R.T., and W. Lick. "Wind Driven Currents in Lake Erie," *J. Geophys. Res.*, 77 (15) (May 1972).

Great Lakes Water Quality Board. Report to the International Joint Commission, Report on Great Lakes Water Quality (1982).

Grotkop, G. "Finite Element Analysis of Long-Period Water Waves," *Comput. Meth. Appl. Mech. Eng.*, 2 (2) (May 1973).

Gupta, S.K., and K.K. Tanji. "Computer Program for Solution of Large, Sparce, Unsymmetric Systems of Linear Equations," *Int. J. Numeric. Meth. Eng.* 11:1251-1259 (1977).

Hamdy, Y., J.D. Kinkead and M. Griffiths. St. Marys River Water Quality Investigations, Great Lakes Surveys Unit, Water Resources Branch, Ontario Ministry of the Environment, Report 1978 (March 1973-74).

Hansen, W. "Theorie zur Errechnung des Wasserstandes und der Stromungen in Randmeeren Nebst Anwendungen," *Tellus*, 8 (3) (August 1956).

Hag, A., W.J. Lick and Y.P. Sheng. "The Time dependent Flow in Large Lakes with Application to Lake Erie," Technical Report, Dept. Earth Science, Case Western Reserve University, Cleveland, Ohio (1974) 212.

Heaps, N.S. "A Two-Dimensional Numerical Sea Model," *Philos. Trans., Ser. A*, 265 (1160) (October 1969).

Hinze, J.O. *Turbulence* (New York: McGraw-Hill Book Company, 1959).

Huyakorn, P.S. "Solution of the Steady State Convective Transport Equation Using an Upwind Finite Element Scheme," in *Applied Mathematical Modelling, Vol. 1* (Stoneham, MA: Butterworth-Heinemann, 1977), p. 187.

Ibrahim, K. "Simulation of Pollutant Transport Responses to Loading and Weather Variations in Lake St. Clair and the Connecting Channels," Ph.D. thesis, University of Windsor, Windsor, Ontario (1986).

Ibrahim, K., and J.A. McCorquodale. "Finite Element Circulation Model for Lake St. Clair," *J. Great Lakes Res. Int. Assoc. Great Lakes,* 11 (3):208-222 (1985).

Imam, E.H. "Numerical Modelling of Rectangular Clarifiers," Ph.D. thesis, University of Windsor, Windsor, Ontario (1981).

Industry Self Monitoring. Industry Self Monitoring Field Programme Ontario Ministry of the Environment, Toronto, Canada, unpublished data, (1986).

Jelesnianski, C.P. "Numerical Computations of Storm Surges with Bottom Stress," *Mon. Weather Rev.* 95:740-756 (1967).

Katopodes, N.D. "A Dissipative Galerkin Scheme for Open Channel Flow," *J. Hydraul. Div., ASCE*, 110 (4):450-466 (1984).

Kennedy, J.B., and A.M. Neville. *Basic Statistical Methods for Engineers and Scientists.* 2nd ed. (New York, NY: Harper and Row, 1976).

Kinnmark, I.P.E., and W.G. Gray. "A Two Dimensional Analysis of the Wave Equation Model for Finite Element Tidal Computations," *Int. J. Numeric. Meth. Eng.* 20:369-383 (1984).

Kizlauskaus, A.G., and P.L. Katz. "A Numerical Model for Summer Flows in Lake Michigan," *Arch. Meteor. Geophys. Bioklim.*, 12:181-197 (1974).

Krauss, W. *Methods and Results of Theoretical Oceanography, Vol. 1,* (Berlin: Gebr. Borntrager, 1973).

Krishnappan, B.G. and Y.L. Lau. RIVMIX Transport Model, computer programme developed by CCIW (1973).

Laible, J.P. "Recent Developments in the Use of the Wave Equation for Finite Element Modeling of Three Dimensional Flow," Department of Civil and Mechanical Engineering, University of Vermont, Burlington, Vermont (1984).

Lau, Y.L., and B.G. Krishnappan. "Modeling Transverse Mixing in Natural Streams," *J. Hydraul. Div., ASCE,* 107:209-226 (1981).

Launder, B.E., and D.B. Spalding. *Mathematical Models of Turbulence* (New York: Academic Press, 1972).

Leedertse, J.J. "Aspects of a Computational Model for Long Period Water Wave Propagation," Memorandum, RM-5294-PR, Rand Corporation (May 1967).

Lorenzon, M.W., C.W. Chen, E.K. Noda, and L.S. Hwang. "Review of Evaluation Methodologies for Lake Erie Wastewater Management Study," Report to the U.S. Army Corps of Engineers District, Buffalo, NY (1974).

Lick, W. "Numerical Models of Lake Currents," U.S. EPA-NTIS No. 600-3-76-020 (1976).

Liggett, J.A. "Unsteady Circulation in Shallow, Homogeneous Lakes," *J. Hydraul. Div. ASCE,* (4) (July 1969).

Liggett, J.A. "Lake Circulation," in *Unsteady Flow in Open Channels, Vol. II,* (Fort Collins, Colorado: Water Resources Publications, 1975).

Liggett, J.A., and K.K. Lee. "Properties of Circulation in Stratified Lakes," *J. Hydraul. Div., ASCE,* (1) (January 1971).

Liggett, J.A., and C. Hadjitheodorou. "Circulation in Shallow Homogeneous Lakes," *J. Hydraul. Div., ASCE,* pp. 609-620 (March 1969).

Liggett, J.A., and D.A. Woolhiser. "Difference Solutions of the Shallow Water Equation," *J. Eng. Mech. Div., ASCE,* (2) (April 1967).

Lyman, W.J., W.F. Reehl and D.H. Rosenblatt. *Handbook of Chemical Property Estimation Methods. Environmental Behaviour of Organic Compounds* (New York: McGraw-Hill, 1982).

Lynch, D.R., and W.G. Gray. "A Wave Equation Model for Finite Element Tidal Computations," in *Computers and Fluids, Vol. 7* (New York: Pergamon Press, 1979), pp. 207-228.

Mackay, D., and P.J. Leinonen. "Rate of Evaporation of Low-Solubility Contaminants from Water Bodies to Atmospheres," *Environ. Sci. Technol.* 7:611-614 (1975).

McCorquodale, J.A., E.H. Imam, J.K. Bewtra, Y.S. Hamdy and J.K. Kinkead. "Transport of Pollutants in Natural Streams," *Can. J. Civil. Eng.* 10 (1):9-17 (1983).

McCorquodale, J.A., and K. Ibrahim. "Application of the K-E Model and TOXIWASP Model to the Determination of HCB Distribution in the St. Clair River Ecosystem," IRI, Ministry of Environment, Toronto, Ontario (1985).

McCorquodale, J.A., K. Ibrahim and Y.S. Hamdy. "Fate and Transport Modelling of Perchloroethylene in the St. Clair River," *Water Poll. Res. J. Can.* 21 (3):398-410 (1986).

McCorquodale, J.A., and E.M. Yuen. "St. Marys River Hydrodynamic and Dispersion Study," A Report Submitted to the Ontario Ministry of Environment, Industrial Research Institute, University of Windsor, Report #IRI 18-61 (1987).

McCorquodale, J.A., and E.M. Yuen. "Report on St. Marys River Modelling for the Upper Great Lakes Connecting Channel Study," A Draft Report to MOE, IRI 20-23 (1988).

Mills, W.B., J.D. Dean, D.B. Porcella, S.A. Gherini, R.J.M. Hudson, W.E. Frick, G.L. Rupp and G.L. Bowie. "Water Quality Assessment: A Screening Procedure for Toxic and Conventional Pollutants, Part 1," U.S. EPA Report-600/6-82-004a (1982), p. 570.

Mills, W.B., J.D. Dean, D.B. Porcella, S.A. Ungs, S.A. Gherini, K.V. Summers, Mok Lingfung, G.L. Rupp, G.L. Bowie and D.A. Haith. "Water Quality Assessment: A Screening Procedure for Toxic and Conventional Pollutants, Parts 1 and 2," U.S. EPA Report-600/6-85-002a and b (1985).

MISA "Municipal Industrial Strategy for Abatement Field Programme," Ontario Ministry of the Environment (1986).

Moins, S.M.A. "Moving Finite Element Solution of Discontinuous Open Channel Flow," Ph.D. thesis, McMaster University, Hamilton, Ontario (1988).

Norton, W.R. et al. "A Finite Element Model for Lower Granite Reservoir," Water Resources Engineers, Inc., Walnut Creek, CA (March 1973).

O'Connor, D.J., J.A. Mueller and K.J. Farley. "Distribution of Kepone in the James River Estuary," *J. Environ. Eng. Div., ASCE,* 109 (2):396-413 (1983).

Ontario Ministry of Environment. "Water Management, Goals, Policies, Objectives and Implementation Procedures of the Ministry of the Environment" (Blue Book) (1984).

Ontario Ministry of the Enviroment. "St. Marys River Water Quality Survey" Report 1986 (1983).

Paul, J.F., and W. Lick. "A Numerical Model of Thermal Plumes and River Discharges," Proceedings of the 17th Conferences of Great Lakes Research, International Association of Great Lakes Research (1974), pp. 445-455.

Phillips, D.W., and J.G. Irbe. "Lake to Land Comparison of Wind, Temperature, and Humidity on Lake Ontario during the International Field Year for the Great Lakes (IFYGL)," Atmos. Env. Service Rept. CLI-2-77. 4905 Dufferin St., Downsview, Ont. M3H 5T4 (1978).

Pinder, G.F., and W.G. Gray. *Finite Element Simulation in Surface and Subsurface Hydrology* (New York: Academic Press, 1977).

Potter, D. *Computational Physics* (New York: John Wiley & Sons Inc., 1973).

Rao, D.B., and T.S. Murty. "Calculations of the Steady-State Wind-Driven Circulations in Lake Ontario," *Arch. Meteorol. Geophys. Bioklim.,* 19:195-210 (1970).

Raudkivi, A.J., and R.A. Callander. *Advanced Fluid Mechanics. An Introduction* (London: Edward Arnold Ltd., 1975), p. 60.

Rastogi, A.K., and W. Rodi. "Predictions of Heat and Mass Transfer in Open Channels," *J. Hydraul. Div., ASCE,* 104 (3):397-420 (1978).

Reid, R.O., and B.R. Bodine. "Numerical Model for Storm Surges in Galveston Bay. *J. Waterways Harbours Div., ASCE,* WW1 (February 1968).

Richardson, W.L., and V.J. Bierman, Jr. "A Mathematical Model of Pollutant Cause and Effect in Saginaw Bay, Lake Huron," Water Quality Criteria Research of the U.S. EPA, EPA-600/3-76-079 (1976).

Richardson, W.L., J.C. Filkins, and R.V. Thomann. "Preliminary Analysis of the Distribution and Mass Balance of PCB in Saginaw Bay," U.S. EPA Report Gross Ile Lab., Grosse Ile, MI (1977).

Richtmyer, R.D., and K.W. Morton. *Difference Methods for Initial-Value Problems* (New York: Interscience Publishers, 1967).

Roach, P.J. *Computational Fluid Dynamics* (Albuquerque, New Mexico: Hermosa Publishers, 1976).

Roesch, S.E., L.J. Clark, and M.M. Bray. "User's Manual for the Dynamnic (Potomac) Estuary Model," U.S. EPA Report-903/9-79-001 (1979).

Rodi, W. *Turbulence Models and Their Application in Hydraulics. A State of the Art Review,* 2nd ed. (Delft, The Netherlands: Intl. Assoc. Hydraulic Research, 1984).

Rodi, W., R.N. Pavlovic and S.K. Srivatsa. "Prediction of Flow and Pollutant Spreading in Rivers," in *Transport Models for Inland and Coastal Waters, Proceedings of a Symposium on Predictive Ability,* H.B. Fisher, Ed., (London: Academic Press, 1981).

Scavia, D., B.J. Eadie and A. Robertson. "An Ecological Model for Lake Ontario, Model Formulation, Calibration, and Preliminary Evaluation," Great Lakes Environmental Research Laboratory, NOAA, Tech. Report ERL-371-GLERL 12 (1976).

Schlichting, H. *Boundary Layer Theory* (New York: McGraw-Hill, 1968).

Schwab, D.J., J.R. Bennett and A.T. Jessup. "A Two Dimensional Lake Circulation Modeling System," NOAA Tech. Memo. ERL, GLERL-38, National Technical Information Service, Springfield, VA, 22161 (1981), p. 79.

Schwab, D.J., and J.A. Morton. "Estimation of Overlake Wind Speed from Overland Wind Speed: a Comparison of Three Methods," *J. Great Lakes Res.* 10:68-72 (1984).

Shirazi, M.A., and L.R. Davis. "Workbook of Thermal Plume Prediction, Vol. 2: Surface Discharge," U.S. EPA Report-R2-72-0056 (1974).

Shen, H.T., P.D. Yapa and M.E. Petroski. "Simulation of Oil Slick Transport in Great Lakes Connecting Channels," Report No. 86-3 Department of Civil and Environmental Engineering, Clarkson University. Potsdam, New York 13676 (March 1986).

Sheng, Y.P. "Wind-Driven Currents and Dispersions of Contaminants in the Near-Shore of Large Lakes," Ph.D. thesis, School of Engineering, Case Western Reserve University, Cleveland, OH (1975).

Sheng, Y.P., and W. Lick. "A Two-Mode Free-Surface Model for the Time Dependent, Three-Dimensional Flows in Large Lakes," U.S. EPA Report (1977), p. 100.

Sheng, Y.P., W. Lick, R.T. Gedney and F.B. Molls. "Numerical Computation of Three-Dimensional Circulation in Lake Erie: A Comparison of a Free-Surface Model and a Rigid Lid Model," *J. Phys. Oceanogr.* 8:713-727 (1978).

Simons, T.J. "Development of Numerical Models of Lake Ontario," Proceedings 15th Conference Great Lakes Research, International Association of Great Lakes Research (1972) pp. 655-672.

Simons, T.J. "Verification of Numerical Models of Lake Ontario, Part I: Circulation in Spring, Early Summer," *J. Phys. Oceanogr.* 4:507-523 (1974).

Simons, T.J. "Verification of Numerical Models of Lake Ontario, Part II: Stratified Circulation and Temperature Changes," *J. Phys. Oceanogr.* 5:98-110 (1975).

Simons, T.J. "Continuous Dynamical Calculations of Water Transports in Lake Erie in 1970," *J. Fish. Res. Bd. Can.* 33:371-384 (1976).

Stolzenbach, K.D., and D.R.F. Harleman. "An Analytical and Experimental Investigation of Surface Discharges of Heated Water," MIT, R.M. Parsons Lab, Report No. 135 (1971).

Stone, H.L., and P.L. Brian. "Numerical Solution of Convective Transport Problems," *Am. Inst. Chem. Eng. J.* 9 (5):681-688 (1963).

Taylor, C., and J. Davis. "Tidal and Long Wave Propagation — A Finite Element Approach," Department of Civil Engineering, University College of Swansea, C/R/189/72 (1972).

Taylor, G.I. "Dispersion of Soluble Matter in Solvent Flowing Slowly Through a Tube," *Proc. Royal Society (London)*, A, 219:186-203 (1953).

Taylor, G.I. "The Dispersion of Matter in Turbulent Flow Through a Pipe," *Proc. Royal Society (London)*, A, 223:446-468 (1954).

Thomann, R.V., D.M. Di Toro, R.D. Winfield and D.J. O'Connor. "Mathematical Modeling of Phytoplankton in Lake Ontario. 1: Model Development and Verification," U.S. EPA Report-660/3-75-005 (1975), p. 177.

Thomann, R.V., and J.A. Mueller. *Principles of Surface Water Quality Modeling and Control* (New York: Harper & Row, 1987).

UGLCC. Upper Great Lakes Connecting Channels Study Field Programme (1986).

UGLCC. Upper Great Lakes Connecting Channels Study, Volumes I & II (1988).

"St. Marys River Oil/Toxic Substance Spill Study Current Velocities and Directions 1980-1983," U.S. Army Corps of Engineers, Detroit District, Great Lakes Hydraulics and Hydrology Branch (December 1984).

"St. Marys River — Soo Harbor and Little Rapids Cut Ice Conditions and Ice Boom Operations," U.S. Army Corps of Engineers (1986).

Welander, P. "Wind Action on a Shallow Sea: Some Generalizations of Ekman's Theory," *Tellus.* 9(1):45-52 (February 1957).

Whitman, R.G. "A Preliminary Confirmation of the Two-Film Theory of Gas Adsorption," *Chem. Metall. Eng.* 29:146-148 (1923).

Yotsukura, N., H.B. Fisher and W.W. Sayre. "Measurement of Mixing Characteristics of the Missiouri River between Sioux City, Iowa, and Plattsmouth, Nebraska," U.S. Geol. Surv. Water Supply Paper 1899-G (1970).

Yotsukura, N., and W.W. Sayre. "Transverse Mixing in Natural Channels," *Water Resour. Res.* 12 (4):695-704 (1976).

Yuen, E.M. "Modelling of Toxic Contaminants in Large Rivers. Ph.D. thesis, University of Windsor, Windsor, Ontario (1988).

Zienkiewicz, O.C. *The Finite Element Method* (London: McGraw-Hill, 1977).

2 HYDRODYNAMIC MODELING IN THE GREAT LAKES FROM 1950 TO 1990 AND PROSPECTS FOR THE 1990S

INTRODUCTION

Hydrodynamic processes in the Great Lakes directly affect the chemical, biological and ecological dynamics of the system. Horizontal and vertical transport and mixing influence the distribution of nutrients, contaminants, and biota. This paper discusses some of the conceptual and numerical hydrodynamic models that have been developed for the Great Lakes. It is not intended to serve as a tutorial on lake hydrodynamics but rather as a brief introduction to some of the different types of hydrodynamic models that have been developed for the Great Lakes and as a reference source for researchers who want to investigate the subject further. Several excellent tutorials and reviews on hydrodynamic modeling in the Great Lakes have already been published by Mortimer (1974 and 1984), Csanady (1984), and Boyce et al. (1989). A book on circulation modeling was written by Simons (1980), and a book on coastal hydrodynamics was published by Csanady (1982). The relation of hydrodynamic modeling to biological and chemical processes is covered specifically in reviews by Boyce (1974), Simons (1976c), and Bedford and Abdelrhman (1987).

Hydrodynamic models that have been developed for the Great Lakes can be categorized either as models dealing mainly with water level fluctuations in the lakes, which need not be particularly concerned with the details of horizontal motion in the lake, or models of lake circulation and thermal structure, which are mainly concerned with subsurface fluid motions and thermodynamics. Models of water level fluctuations include storm surge models, models of seiches and normal modes, tidal models, wind wave models, and hydrologic models. The main types of lake circulation and thermal structure models are those dealing only with the horizontal motions, and those that incorporate horizontal and

ISBN 0-87371-511-X

Table 1.   Physical Parameters Involved in Models
of Water Level Fluctuations

Type of Model	Physical Parameters
Tides	Tidal force, surface slope, bottom friction, Coriolis force
Storm surge	Wind stress, pressure gradient, water surface slope, bottom friction, Coriolis force
Seiches and normal modes	Inertia, bottom friction, Coriolis force
Wind waves	Wind stress, wave energy, wave dissipation
Hydrologic models	Precipitation, runoff, evaporation, water level

vertical motion and thermal structure, physical (scale) models, and water quality models. This paper will describe the primary physical parameters involved in each of these types of models and provide references to some of the more important developments in each area.

WATER LEVEL FLUCTUATIONS

One of the first physical phenomena to be recognized as a significant hydrodynamic process in the Great Lakes was the elevation or depression of the water level at the shoreline. These fluctuations can be caused by astronomical forces (tides), by the force of storm winds or atmospheric pressure disturbances (storm surges and wind waves), or by the periodic oscillation of the lake surface after storm forcing has ceased (seiches). From a modeling standpoint, the primary dynamical variable in these processes is the water level fluctuation at each point in the lake. The physical parameters involved in modeling these phenomena are listed in Table 1.

TIDES

The dominant tides in the Great Lakes are the 12.42 h lunar and 12.00 h solar semidiurnal tides (M2 and S2). Diurnal water level oscillations have also been observed, but these are mainly a result of diurnal oscillations in the overlake wind field. The maximum range of the tidal oscillation is generally less than 10 cm. A pronounced resonance with the natural free oscillation period in Green Bay results in tidal ranges that can reach 18 cm at the head of the Bay. As described by Mortimer (1965), a Jesuit missionary named Father Louis Andre may have provided the earliest observations of tides in the Great Lakes in a 1676 report from Green Bay. As cited by Defant (1961), Harris (1907) a report that estimated the amplitude (half the range) of the combined lunar and solar semidiurnal tides as 1.1 cm at Milwaukee on Lake Michigan, 0.7 cm at Marquette and 3 cm at Duluth on Lake Superior. Endros (1930) used the data of Henry (1902) to estimate an amplitude of 7.1 cm for the combined tide at Amsterburg at the west end of Lake Erie. Platzman (1966) showed how diurnal oscillations of the wind over Lake Erie produce a "wind tide" oscillation in the water levels with an average amplitude of 1.5 cm at the ends of the basin. The rotation of the earth imparts a rotary character to the tides in the Great Lakes so that the high tide progresses around the shore of the lake in either a clockwise or counterclockwise direction. Mortimer and Fee (1976) describe the phase progression of the semidiurnal tides for Lakes Michigan and Superior. Hamblin (1976) discusses the theoretical basis for the sense of rotation of tides in the lakes. Hamblin (1987) includes a section on the tidal response of Lake Erie.

Storm Surges

When a steady wind blows along a channel, the equilibrium condition of the water surface in the channel is a depression of the water level on the upwind end and an elevation of the water level on the downwind end. In a channel of uniform depth, the magnitude of the depression and elevation are the same and are proportional to the length of the channel, the square of the wind speed, and the inverse of the depth. Early studies of storm surges on the lakes such as Keulegan (1951) used this equilibrium condition to calculate water level deviations for given wind speeds. In some of the first models of storm surges on the Great Lakes, Hayford (1922), Keulegan (1953), and Hunt (1959) took account of variations in lake depth by segmenting the lake along its axis and calculating equilibrium solutions for each segment. The results of these studies were good when the prevailing wind could be approximated as a steady, uniform wind blowing along the main axis of the lake.

The hydrodynamics of storm surges in the Great Lakes are governed by the mass and momentum conservation equations of shallow water theory. The space- and time-dependent wind stress is the upper boundary condition. The equilibrium state of a channel is just a special case for which an analytic solution is possible. In general, time-dependent equations applied to a two-dimensional lake of arbitrary shape have no analytic solution, but can be solved approximately on a computer by applying them in a finite difference form on a mesh of points covering the lake. Platzman (1958, 1963) performed such calculations for moving pressure disturbances on Lake Michigan and for nine actual storm surge cases on Lake Erie. He showed how the earth's rotation causes the point of maximum water level displacement to progress counterclockwise around the edge of the lake. The results obtained for a moving pressure disturbance cannot be duplicated by an equilibrium model and the results for actual Lake Erie storm surges showed considerable improvement over the equilibrium method.

A linear dynamic model was applied to fifteen Lake Erie storm surge cases by Schwab (1978). In addition to the time-dependence and two-dimensionality of the windfield, this study also included the effect of atmospheric stability and wind speed on drag coefficient and land-lake wind speed ratio. These factors are significant in many storm surge episodes. The models of Schwab (1978) and Platzman (1965) now form the basis for routine operational storm surge forecast systems for Lakes Erie and Michigan respectively. Regression models, which develop a statistical regression relationship between storm surges and meteorological forcing, have also been used successfully in predicting storm surges, particularly on Lake Erie, by Harris and Angelo (1963) and Richardson and Pore (1969).

Several detailed case studies of storm surges have been published which provide insight into the meteorological conditions that are conducive to storm surges on the Great Lakes. Irish and Platzman (1962) examine and categorize the types of storms that generally cause storm surges on Lake Erie. Ewing et al. (1954), Donn (1959), Freeman and Murty (1972), Murty and Polavarapu (1975), Hamblin (1979), and Dingman and Bedford (1984) describe several specific episodes of extreme surges on Lakes Michigan, Huron, Ontario, and Erie.

Schwab (1978), Budgell and El-Shaarawi (1979), and Simons and Schertzer (1989) describe how the inherently linear dynamics of storm surges in the Great Lakes are amenable to solution by the impulse response function method. This technique represents the time-dependent water level fluctuation at a point on the lake as the superposition of responses to a series of impulsive wind stresses that approximate the continuously changing actual wind. Schwab (1982) showed further that this method could be inverted to provide an estimation of overlake wind fields from observations of water level fluctuations around the shore of the lake. Heaps et al. (1982) and Freeman et al. (1974) explore the problem of coupling the storm surge response of bays and harbors to the open lake. Hamblin (1987) provides a summary of these techniques and other research that has been done on Lake Erie storm surges.

Seiches

Because of the frequent passage of extra-tropical storms through the Great Lakes region, the lakes are often subject to sustained strong winds. The winds tend to push the water in the lake to the downwind end, elevating the water level there and depressing the water level on the upwind end. When the wind diminishes, the tilted water surface tends to return to its normal position, but sometimes oscillates several times before returning to normal. The subsequent oscillation of the water surface of the lake after a strong wind has subsided or after an atmospheric pressure disturbance has caused a water level disturbance to develop in the lake is called a seiche. Each lake has its own characteristic period of oscillation for the fundamental (unimodal) seiche and higher harmonics. The period of oscillation depends on the size of the lake and its mean depth. The simplest models of seiches assume that the lake can be approximated as a channel, with seiche motions confined to the longitudinal axis of the channel. Harris (1953) and Defant (1953) discuss the hydrodynamic theory of one-dimensional, or channel, seiche models. Platzman and Rao (1964a, 1964b) describe a comprehensive application of a one-dimensional numerical seiche model to lake Erie and comparisons with observations of periodic water level fluctuations at standard water level gaging stations around the lake. Mortimer (1965) applied the channel theory to Lake Michigan proper and to Green Bay separately to show the relation of seiche periods in the bay and in the lake to tidal periods. Rockwell (1966) systematically calculated seiche periods using a one-dimensional numerical model for all five lakes.

Some lakes, particularly Superior and Huron, have very complicated shorelines that cannot be simply approximated as a channel. In addition, the effects of the earth's rotation on seiching motions (which can significantly alter the structure of the longer period modes) cannot be fully accounted for in a one-dimensional channel model. Seiche, or normal mode, models based on the hydrodynamics in a fully two-dimensional rotating basin were developed by Hamblin (1972, 1987), Platzman (1972), Rao and Schwab (1976), Rao et al. (1976), and Schwab and Rao (1977). These models have been able to accurately depict the two-dimensional structure and amplitude of lake surface oscillations for all the lakes. The results of these models have been used to explain the dynamical response of the lakes to storm surges and tidal forces, particularly the rotational charac-teristics.

Wind Waves

The development of wind waves on the Great Lakes is governed by the same physical parameters as ocean wind waves, namely a balance between wind energy input, wave energy, and wave dissipation. The main distinguishing feature of lake wind waves is that they are generally locally generated, not propagated from a distance, as is the case for many large ocean waves. Therefore, many of the techniques used to model ocean wave generation can be successfully applied to lake waves. The simplest wave models consist of a single equation relating wave height at a particular place and time to an empirical function (sometimes a complicated mathematical formula) of wind speed and fetch distance. Another equation is used to calculate wave period. The evaluation of these equations can usually be carried out with a pocket calculator or looked up on a graph. The SMB (Sverdrup, Munk, Bretschneider-Bretschneider, 1970 and 1973) model has been widely used for ocean wave forecasting and is the basis of a method for automated operational Great Lakes wave forecasts (Pore, 1979). A one-dimensional model developed by Donelan (1980) can also predict the direction from which maximum wave energy will arrive at the forecast point, which may not be coincident with the wind direction. Bishop (1983) compared predictions from three different one-dimensional methods for wave forecasting to observations of wave height and wave period in Lake Ontario. His results showed that the Donelan method was slightly better than the ocean wave methods, particularly when the dominant wave direction differed from the wind direction. Two-dimensional models for wave prediction attempt to predict waves at all points in the lake (or region of the ocean) at the same time. Actually, waves are not predicted at every point, but rather the lake (or region of the ocean) is split into a very large number of subregions and average wave conditions are predicted for each subregion. The difference from one-dimensional prediction methods is that in two-dimensional methods the results for any one subregion depend not only on wind speed and fetch distance, but also on the energy fluxes from surrounding subregions. That is, a budget of wave energy is maintained for each subregion which includes energy input from the wind, energy propagated into this subregion from surrounding regions, and energy propagated out of the region to surrounding regions.

A two-dimensional wave prediction model for the Great Lakes has been developed by Schwab et al. (1984) and Liu et al. (1984). Given a description of the lake topography and the two-dimensional, time-dependent wind field over the lake, the model predicts wave height, wave period, and wave direction for an array of square grid boxes covering the lake. The model has been successfully tested against one-dimensional methods for steady conditions in ideal basins and against several sets of actual observations of wave height, period, and direction in the Great Lakes.

Hydrologic Models

Hydrologic models are used to calculate the balance between precipitation, runoff, inflow, outflow, evaporation, and changes in the mean lake level. In the Great lakes, manmade control structures such as locks and dams, control some of the inflows and outflows. Since the lakes are a single, connected hydrologic system, changes in the water balance of the upstream lakes affect all of the lakes downstream. Table 2 (Derecki, 1976) shows the relative magnitudes of the terms in the water balance equation for the four hydrologic lake basins.

If values for some of the terms in the hydrologic balance equation are known accurately enough, it is possible to develop models to estimate the other terms. For example, observed water level changes, inflow, outflow, precipitation, and runoff can be used to estimate evaporation. During the 1972 International Field Year on the Great Lakes (IFYGL) experiment, an attempt was made to measure each of the terms in the water balance for Lake Ontario as precisely as possible. In the IFYGL summary (Aubert and Richards 1981), chapters on Meteorology, Precipitation, Atmospheric Water Balance, Energy Balance, Terrestrial Water Balance, and Evaporation Synthesis detail the methods and results.

A comprehensive operational hydrologic model for Lakes Michigan, Huron, St. Clair, and Erie, and their connecting channels is described by Quinn (1978). Another model for forecasting lake levels up to six months in advance was

Table 2.   Average Hydrologic Water Budget (cm) for the Great
Lakes, 1937–1969 (Derecki, 1976)

Lake	P	R	I	O	E
Superior	80	58	0	86	55
Michigan-Huron	80	67	60	139	65
Erie	88	72	640	706	85
Ontario	84	150	927	1077	70

Note: P = precipitation, R = runoff, I = inflow, O = outflow, E = evaporation.

developed by Croley and Hartmann (1986). These types of models are now used routinely by the U.S. Army Corps of Engineers and the International Joint Commission to develop lake level regulation and water management plans.

LAKE CIRCULATION AND THERMAL STRUCTURE

Currents in the lakes are the result of three main forcing mechanisms, namely, hydraulic (river) flow, wind forcing, and thermal forcing (heating by the sun). Because the ratio of horizontal scale (~100km) to vertical scale (~100m) for the lakes is so large, horizontal motions predominate over vertical motions. Models of lake circulation can be categorized as (1) those dealing primarily with the horizontal circulation due to hydraulic forces and wind stress, (2) those that include the effects of thermal forcing and vertical stratification, (3) physical models (scale models), and (4) water quality models that are more concerned with the effect of circulation on the advection, diffusion, and ultimate distribution of suspended or dissolved substances in the lakes. The primary physical parameters involved in these models are listed in Table 3.

Horizontal Circulation

In shallow lakes with sloping bottoms, a steady wind stress over the lake pushes surface water downwind. The water level at the downwind end of the lake rises and the resultant pressure gradient causes an upwind return flow in the deeper part of the lake. The first order force balance is between the wind stress and the water level slope, and a mass balance is established between downwind surface current and upwind return flow in the deeper part of the lake. In a lake with a sloping bottom, this results in strong currents in the direction of the wind in the nearshore region and weaker upwind currents in the deeper parts of the lake. This so called "two-gyre" pattern with a clockwise circulation cell on the left side of the wind direction and a counterclockwise cell to the right is the dominant response of the lake to steady wind forcing. When the wind stress diminishes or ceases, the rotation of the earth causes the gyre pattern to rotate

Table 3.   Physical Parameters Involved in Models of
Lake Circulation and Thermal Structure

Type of Model	Physical Parameters
Horizontal circulation	Wind stress, bottom topography, bottom friction, internal shear stresses, Coriolis force
Three-dimensional circulation and thermal structure	Wind stress, heat flux, internal pressure gradients, bottom topography, bottom friction, internal shear stresses, Coriolis force, buoyancy
Physical models (scale models)	Wind stress, bottom topography, bottom friction, Coriolis force, scale effects
Water quality models	Advection, diffusion, partition between dissolved and suspended states, reaction rates, deposition, resuspension

counterclockwise around the basin with a characteristic period that depends on the specific geometry of the basin and the latitude. For a 2×1 elliptic paraboloid at 45°N this period is about 5 days. Conceptual models of the horizontal circulation in the lakes due to steady wind forcing were developed by Birchfield (1967, 1969, 1972a, 1972b), Csanady (1967, 1968b, 1973b, 1976b), Bennett (1974), Thomas (1975), and Lien and Hoopes (1978). Numerical models which incorporated the particular geometry of a specific lake and could make predictions of circulation patterns for different wind directions were developed by Hamblin (1969), Murty and Rao (1970), Rao and Murty (1970), Freeman et al. (1972), Bonham-Carter and Thomas (1973), and Gallagher et al. (1973). Sheng et al. (1978) compare the results of a model of Lake Erie that includes free surface fluctuations as one of the dynamic variables to results from one that does not. They found that for changes in lake circulation over periods longer than about half a day, free surface fluctuations were not an important factor. Pickett (1980) showed that this type of model was able to explain many of the observed characteristics of wintertime circulation in all five lakes.

If an assumption is made about the distribution of eddy viscosity in the vertical direction, the vertical distribution of currents can also be estimated in the steady-state pattern. Models by Gedney and Lick (1972) and Witten and Thomas (1976) are good examples. Of course in natural conditions, the wind over a lake cannot usually be approximated as a steady wind so that time-dependent changes in the wind must also be taken into account. The time-dependent response of lake circulation to wind stress was treated in numerical models by Liggett (1969,

1970), Liggett and Hadjitheodorou (1969), Birchfield and Murty (1974) and analyzed mathematically by Birchfield and Hickie (1977). The time-dependent response of the lake is linked to the characteristic rotational mode of the lake as discussed above. These modes were noted in the work of Rao and Schwab (1976) on seiches, and modeled in more detail by Csanady (1976a), Saylor et al. (1980), Bennett and Schwab (1981), Huang and Saylor (1982), and Schwab (1983). The work of Simons (1983, 1984, 1985, 1986) demonstrates the dependence of the response of the lake to wind stress on the topographic modes. He also shows that the time-averaged mean circulation pattern in the lake depends on the rectified effects of nonlinear topographic wave interactions. This explains the difficulty linear models have in reproducing the details of observed long term circulation patterns.

Three Dimensional Circulation and Thermal Structure

The Great Lakes undergo an annual cycle of heating and cooling that typically takes the mean temperature of the lake both above and below the temperature of maximum density for freshwater (close to 4°C). When water temperatures in the lake are near the temperature of maximum density (usually in the spring and fall, sometimes throughout the winter), the forces of buoyancy and internal horizontal pressure gradients are small and generally do not have a significant effect on horizontal circulation patterns. During the period of the year when the net radiation flux to the lake is positive, typically February through September, a balance is maintained in the water column between vertical mixing due to surface wind stress and buoyancy of the warmer surface water. Wind-induced mixing tends to distribute temperature uniformly throughout the water column while the buoyancy forces tend to establish a vertical gradient with warmer water at the surface and cooler water at depth. The result is the development of a surface-mixed layer of warm water separated from a deep layer of cold water by a thermal transition zone whose depth and thickness depend on the relative magnitudes of mixing and buoyancy forces. These processes were observed by Church (1943, 1945) and Millar (1952). Ayers (1956) and Ayers and Bachman (1957) modeled the circulation patterns associated with thermal gradients as a simple geostrophic balance between pressure gradients and Coriolis force. The thermal transition zone, or thermocline, acts as a barrier to vertical mixing of mass or momentum between the upper mixed layer and the lower layer. Models of the development of the surface mixed layer and vertical thermal structure are described by Ivey and Boyce (1982), Ivey and Patterson (1984), Lam and Schertzer (1987), Schertzer et al. (1987), and McCormick and Meadows (1988) whose results point out the limitations of models that assume horizontal homogeneity. Gorham and Boyce (1989) show how the depth of the mixed layer in a lake depends on lake surface area and maximum depth.

In lakes with sloping bottoms, springtime heating warms nearshore shallow water faster than deep water, and a thermal barrier or bar develops which separates warm nearshore water from cooler offshore water. The location of the bar progresses offshore until warm water covers the entire surface of the lake and becomes the upper-mixed layer. Hydrodynamic models of this process were developed by Rodgers (1965, 1966), Scott and Lansing (1967), Csanady (1968a), Elliot (1971), Huang (1971), and Bennett (1971).

During the summer stratification, there is generally a sufficient density gradient between the upper and lower layers to allow internal waves to develop. For internal waves, volume transport in the upper layer at any point in the lake is very nearly compensated for by an equal and opposite volume transport below the thermocline so that large fluctuations of the thermocline can occur without a noticeable change in the free surface level. Internal seiches are generally of longer period than free surface seiches, and therefore are influenced more strongly by the earth's rotation. Observations of internal wave motions in Lake Michigan Lakes were made by Verber (1964, 1966) and interpreted as Kelvin and Poincaré type waves by Mortimer (1963, 1968). Similar observations in Lake Ontario (Boyce and Mortimer 1978), and Lake Erie (Boyce and Chiocchio 1987) demonstrate the ubiquitous nature of internal waves. Internal waves were also observed during a period of wintertime stratification (mixed layer temperatures less the temperature of maximum density) in Lake Ontario by Marmorino (1978). Schwab (1977) developed a numerical hydrodynamic model to calculate the structure of the internal modes of oscillation in Lake Ontario.

The first numerical models of three-dimensional lake circulation approximated the vertical structure of the lake as the superposition of two or more horizontal layers with either permeable or impermeable interfaces. Models of internal normal modes generally incorporate two layers with an impermeable interface. Surface-mixed layer models on the other hand require that heat and momentum can be freely exchanged between layers. Lee and Liggett (1970), Liggett (1970), Liggett and Lee (1971), Bennett (1971, 1977, 1978), Simons (1971, 1972, 1973, 1974, 1975, 1976a, 1976b), Gedney et al. (1973), Kizlauskas and Katz (1973), Allender (1977), and Allender and Saylor (1979) used layered models of varying degrees of sophistication to simulate three-dimensional circulation and thermal structure in the lakes. Bennett and Lindstrom (1977) showed how a simple empirical model of thermocline oscillations in the coastal boundary layer of Lake Ontario could be an effective tool for predicting the response of the nearshore thermocline to wind stress events. Csanady (1971, 1972, 1973a), Bennett (1973, 1975), and Simons (1979) contributed considerably to the understanding of the limitations of this type of model by considering analytic or semi-analytic solutions for several idealized cases.

Physical (Scale) Models

Physical scale models have been used successfully to simulate large scale hydrodynamic properties in rivers, reservoirs, and embayments. The key to making accurate simulations is to attain dynamic similarity of all important physical processes between the scale model and the prototype. For the Great Lakes, three considerable obstacles stand in the way of developing successful physical models. First, the large ratio of horizontal scale to vertical scale of the lakes (~1000:1) is difficult to duplicate in the laboratory, so that vertically distorted models must be used. Second, the processes of turbulent transfer of energy from wind to water and of frictional dissipation of energy at the lake bottom are not well understood and are difficult to simulate in a scale model. Third, the effect of the earth's rotation on circulation in the lakes must be incorporated into the scale model. Despite these obstacles, several scale models have been constructed. For example, Harleman et al. (1964) built a model of Lake Michigan with a horizontal scale of 1:500,000 and a vertical scale of 1:1000. Infrared lamps were used to simulate solar heating and a fan to simulate wind stress. The model was operated on a rotating (7.56 rpm) platform. Aluminum powder was used to track surface currents. Circulation patterns that developed in the scale model for different combinations of thermal forcing and wind stress were similar to patterns produced by numerical models and infrared from direct observations. Physical models were also developed for Lake St. Clair by Ayres (1964), for Lake Erie by Rumer and Robson (1968), Buechi and Rumer (1969), and Howell et al. (1970), and for Lake Ontario by Li et al. (1975).

Water Quality Models

One of the principal reasons for developing models of lake hydrodynamics is to use the circulation patterns predicted by the model to better predict the transport of suspended and dissolved chemical and biological material that can affect water quality. Particulate matter can enter the water column from atmospheric deposition, from river inflow, or from resuspension of benthic material. Once the material is in the water column, some of it may dissolve and some may remain as particulates. In either case, the advection and diffusion of the material are governed by hydrodynamic processes. For particulate material, deposition and resuspension are also controlled by hydrodynamics.

In order to model the distribution of chemical and biological material in the lake, currents and diffusion coefficients determined from hydrodynamic models must be incorporated into the advection terms of the diffusion equation along with appropriate boundary and initial conditions for the substance(s) being modeled. If the substances are reactive, equations governing reaction rates and products must also be incorporated into the model. Thomann et al. (1981) discuss the utility and limitations of comprehensive water quality and ecosystems models using models developed for the 1972 International Field Year on the Great Lakes as examples. These models include the models of Chen et al. (1975), Thomann et al. (1977), Robertson and Scavia (1979), Scavia (1979, 1980), Simons (1976c), Simons and Lam (1980), and a two-dimensional cross section model by Scavia and Bennett (1980).

Some simpler models of advection and diffusion of conservative elements (usually chloride) over long time scales have successfully reproduced observed average concentration patterns in Lake Erie (Boyce and Hamblin 1975, Lam and Simons 1976) and Lake Superior (Lam 1978). Pickett and Dossett (1979) applied a similar advection-diffusion model to the distribution of mirex in Lake Ontario. The models developed by Paul and Lick (1974) for thermal plumes and river discharges in Lake Erie and by Murthy et al. (1986) for pollutant transport along the north shore of Lake Ontario also belong to this category. Csanady (1970) discusses the general characteristics of this type of model.

In western Lake Erie, periodic occurrences of anoxic conditions near the shallow bottom have been a recurring threat to water quality. Models have been used to evaluate hypotheses about the causes of anoxia there (Lam et al. 1983, DiToro et al. 1987, Snodgrass 1987, Lam and Schertzer 1987, Lam et al. 1987a and 1987b) with emphasis on the role of vertical exchange processes and the timing and intensity of thermal stratification.

Many toxic materials attach to particulate matter and are deposited either permanently or temporarily in the lake sediments. Models of sediment resuspension in the bottom boundary layer have been studied by Sheng and Lick (1979), Lee et al. (1981), Bedford and Abdelrhman (1987), and Lesht and Hawley (1987). For tracking oil spills or conservative tracers, a simple particle trajectory model based on calculated horizontal or three-dimensional circulation patterns can be used. This type of model was investigated by Bennett and Clites (1987), Schwab and Bennett (1987), and Schwab et al. (1980) and used for operational oil spill trajectory predictions in models by Simons et al. (1975) and Schwab et al. (1984a).

PROSPECTS FOR THE FUTURE

The contributions that lake hydrodynamic models have made to models of chemical, biological, and ecological dynamics include incorporation of horizontal transport processes in models of the dynamics of dissolved substances, the effect of the springtime nearshore circulation structure on biological activity, and the effect of lake scale circulation on the distribution of toxic chemicals in sediments. Some of the prospects for new contributions to our understanding of environmental fate processes in lakes that can be made by hydrodynamic models in the next decade are improved models of monthly and seasonal mean circulation, models of frontal dynamics, coupling of sedimentation and resuspension models with lake circulation models, coupling of chemical and biological dynamics with lake circulation models, and the prospects for routine operational lake circulation and thermal structure forecasting. Some of the key remaining problems in hydrodynamic modeling of the Great Lakes include adequate specification of boundary conditions and forcing functions, modeling of subtle dynamical balances that can occur in time-averaged flows, and the impact of nonlinear processes and hydrodynamic instability on parameterizations of turbulence and diffusion in numerical models.

REFERENCES

Allender, J.H. "Comparison of Model and Observed Currents in Lake Michigan," *J. Phys. Oceanogr.* 7:711-718 (1977).

Allender, J.H., and J. H. Saylor. "Model and Observed Circulation Throughout the Annual Temperature Cycle of Lake Michigan," *J. Phys. Oceanogr.* 9:573-579 (1979).

Aubert, E.J., and T. L. Richards, Eds. *IFYGL - The International Field Year for the Great Lakes* (Ann Arbor, MI: Natl. Oceanic and Atmos. Admin., Great Lakes Env. Res. Lab., 1981).

Ayers, J.C. "A Dynamic Height Method for the Determination of Currents in Deep Lakes," *Limnol. Oceanogr.* 1:150-161 (1956).

Ayers, J.C. "Currents and Related Problems at Metropolitan Beach, Lake St. Clair," University of Michigan, Great Lakes Res. Div., Spec. Rep. No. 20 (1964).

Ayers, J.C., and R. Bachmann. "Simplified Computations for the Dynamic Height Method of Current Determination in Lakes," *Limnol. Oceanogr.*, 2:155-157 (1957).

Bedford, K.W., and M. Abdelrhman. "Analytical and Experimental Studies of the Benthic Boundary Layer and Their Applicability to Near Bottom Transport in Lake Erie," *J. Great Lakes Res.* 13:628-648 (1987).

Bennett, J.R. "Thermally Driven Lake Currents during the Spring and Fall Transition Periods," Proc. 14th Conf. Great Lakes Res. Int. Assoc. Great Lakes Res. (1971), pp. 535-544.

Bennett, J.R. "A Theory of Large-Amplitude Kelvin Waves," *J. Phys. Oceanogr.* 3:57-60 (1973).

Bennett, J.R. "On the Dynamics of Wind-Driven Lake Currents," *J. Phys. Oceanogr.* 4:400-414 (1974).

Bennett, J.R. "Another Explanation of the Observed Cyclonic Circulation of Large Lakes," *Limnol. Oceanogr.* 20:108-110 (1975).

Bennett, J.R. "A Three-Dimensional Model of Lake Ontario's Summer Circulation: I. Comparison with Observations," *J. Phys. Oceanogr.* 7:591-601 (1977).

Bennett, J.R. "A Three-Dimensional Model of Lake Ontario's Summer Circulation: II. A Diagnostic Study," J. Phys. Oceanogr. 8:1095-1103 (1978).

Bennett, J.R., and E.J. Lindstrom. "A Simple Model of Lake Ontario's Coastal Boundary Layer," *J. Phys. Oceanogr.* 7:620-625 (1977).

Bennett, J.R., and D.J. Schwab. "Calculation of the Rotational Normal Modes of Oceans and Lakes with General Orthogonal Coordinates," *J. Comput. Phys.* 44:359-376 (1981).

Bennett, J.R., and A.H. Clites. "Accuracy of Trajectory Calculation in a Finite-Difference Circulation Model," *J. Comput. Phys.*, 68:272-282 (1987).

Birchfield, G.E. "Horizontal Transport in a Rotating Basin of Parabolic Depth Profile," *J. Geophys. Res.* 72:6155-6163 (1967).

Birchfield, G.E. "Response of a Circular Model Great Lake to a Suddenly Imposed Wind Stress," *J. Geophys. Res.* 74:5547-5554 (1969).

Birchfield, G.E. "Wind-Driven Currents in a Large Lake or Sea," *Arch. Meteorol. Geophys. Bioklimatol.* A21:419-430 (1972a).

Birchfield, G.E. "Theoretical Aspects of Wind-Driven Currents in a Sea or Lake of Variable Depth with no Horizontal Mixing," *J. Phys. Oceanogr.* 2:355-362 (1972b).

Birchfield, G.E., and T.S. Murty. "A Numerical Model for Wind-Driven Circulation in Lakes Michigan and Huron," *Mon. Weather Rev.* 102:157-165 (1974).

Birchfield, G.E., and B.P. Hickie. "The Time-Dependent Response of a Circular Basin of Variable Depth to a Wind Stress," *J. Phys. Oceanogr.* 7:691-701 (1977).

Bishop, C.T. "Comparison of Manual Wave Prediction Models," *J. of Waterway, Port, Coastal and Ocean Eng.* 109(1):1-17 (1983).

Bonham-Carter, G., and J.H. Thomas. "Numerical Calculation of Steady Wind-Driven Currents in Lake Ontario and the Rochester Embayment," Proc. 16th Conf. Great Lakes Res. Int. Assoc. Great Lakes Res. (1973), pp. 640-662.

Boyce, F.M. "Some Aspects of Great Lakes Physics of Importance to Biological and Chemical Processes," *J. Fish. Res. Board Can.* 31:689-730 (1974).

Boyce, F.M., and P.F. Hamblin. "A Simple Diffusion Model of the Mean Field Distribution of Soluble Materials in the Great Lakes," *Limnol. Oceanogr.* 20:511-517 (1975).

Boyce, F.M., and C.H. Mortimer. "IFYGL Temperature Transects, Lake Ontario, 1972," *Dep. Environ. Ottawa. Tech. Bull.* 100:315 (1978).

Boyce, F.M., and F. Chiocchio. "Inertial Frequency Current Oscillations in the Central Basin of Lake Erie," *J. Great Lakes Res.* 13:542-558 (1987).

Boyce, F.M., M.A. Donelan, P.F. Hamblin, C.R. Murthy and T.A. Simons. "Thermal Structure and Circulation in the Great Lakes," *Atmos.-Ocean* 27(4):607-642 (1989).

Bretschneider, C.L. "Wave Forecasting Relations for Wave Generation," *Look Lab, HI.* 1(3) (1970).

Bretschneider, C.L. "Prediction of Waves and Currents," *Look Lab, HI* 3:1-17 (1973).

Budgell, W.P., and A. El-Shaarawi. "Time Series Modelling of Storm Surges in a Medium-Sized Lake". *Predictability and Modelling in Ocean Hydrodynamics," Elsevier Oceanogr. Ser.* 25:197-218 (1979).

Buechi, P.J., and R.R. Rumer. "Wind Induced Circulation Pattern in a Rotating Model of Lake Erie," Proc. 12th Conf. Great Lakes Res., Int. Assoc. Great Lakes Res. (1969), pp. 406-414.

Chen, C.W., M. Lorenzen and D.J. Smith. "A Comprehensive Water Quality-Ecological Model for Lake Ontario," Report No. TC-435, Tetra Tech, Lafayette, CA (1975).

Church, P.E. "The Annual Temperature Cycle of Lake Michigan, I. Cooling from Late Autumn to the Terminal Point, 1941-42," Univ. Chicago Inst. Meteorol. Misc. Rep. 4. (1942).

Church, P.E. "The Annual Temperature Cycle of Lake Michigan. II. Spring Warming and Summer Stationary Periods, 1942," Univ. Chicago. Inst. Meteorol. Misc. Rep. 18 (1945).

Croley, T.E., and H.C. Hartmann. "Near Real-Time Forecasting of Large Lake Water Supplies; A Users Manual," NOAA Technical Memo. ERL GLERL-70, NTIS 22161 (1986).

Csanady, G.T. "Large-Scale Motion in the Great Lakes," *J. Geophys. Res.* 72:4151-4162 (1967).

Csanady, G.T. "Wind-Driven Summer Circulation in the Great Lakes," *J. Geophys. Res.* 73:2579-2589 (1968a).

Csanady, G.T. "Motions in a Great Lake due to a Suddenly Imposed Wind," *J. Geophys. Res.* 73:6435-6447 (1986b).

Csanady, G.T. "Dispersal of Effluents in the Great Lakes," *Water Res.* 4:79-114 (1970).

Csanady, G.T. "Baroclinic Boundary Currents and Long-Edge Waves in Basins with Sloping Shores," *J. Phys. Oceanogr.* 1:92-104 (1971).

Csanady, G.T. "Response of Large Stratified Lakes to Wind," *J. Phys. Oceanogr.* 2:3-13 (1972).

Csanady, G.T. "Transverse Internal Seiches in Large Oblong Lakes and Marginal Seas," *J. Phys. Oceanogr.* 3:439-447 (1973a).

Csanady, G.T. "Wind-Induced Barotropic Motions in Long Lakes," *J. Phys. Oceanogr.* 4:357-371 (1973b)

Csanady, G.T. "Topographic Waves in Lake Ontario," *J. Phys. Oceanogr.* 6:93-103 (1976a)

Csanady, G.T. "Mean Circulation in Shallow Seas," *J. Geophys. Res.* 81:5389-5399 (1976b).

Csanady, G.T. "The Arrested Topographic Wave," *J. Phys. Oceanogr.* 8:47-62 (1978).

Csanady, G.T. *Circulation in the Coastal Ocean* (Dordrecht, Holland: D. Reidel Publ. Co., 1982).

Csanady, G.T. "Milestones of Research on the Physical Limnology of the Great Lakes," *J. Great Lakes Res.* 10:114-125 (1984).

Defant, A. *Physical Oceanography, Vol. II* (London: Pergamon Press, 1961).

Defant, F. "Theorie der Seiches des Michigansees und ihre Abwandlung durch Wirkung der Corioliskraft," *Arch. Met. Geophys. Bioklimatol. Wien.* A6:218-241 (1953).

Derecki, J.A. "Hydrometeorology: Climate and Hydrology of the Great Lakes. Appendix 4, Limnology of Lakes and Embayments," Great Lakes Basin Framework Study, Great Lakes Basin Commission, Ann Arbor, MI. (1976), pp. 71-104.

Dingman, J.S., and K.W. Bedford "The Lake Erie Response to the January 26, 1978, Cyclone," *J. Geophys. Res.* 89:6427-6445 (1984).

DiToro, D.M. "Vertical Interactions in Phytoplankton Populations — An Asymptotic Eigenvalue Analysis (IFYGL)," Proc. 17th Conf. Great Lakes Res., Int. Assoc. Great Lakes Res. (1974), pp. 17-27.

Donelan, M.A. "Similarity Theory Applied to the Forecasting of Wave Heights, Periods, and Directions," Proc. of the Can. Coastal Conf., Nat. Res. Council, Canada (1980), pp .47-61.

Donn, W.L. "The Great Lakes Storm Surge of May 5, 1952," *J. Geophys. Res.* 64:191-198 (1959).

Elliot, G.H. "A Mathematical Study of the Thermal Bar," Proc. 14th Conf. Great Lakes Res., Int. Assoc. Great Lakes Res. (1971), pp.545-554.

Endros, A. "Gezeitenbeobachtungen in Binnenseen," *Ann. Hydr. Mar. Meteorol.*, 58:305 (1930).

Ewing, M., F. Press and W.L. Donn. "An Explanation of the Lake Michigan Surge of June 26, 1954," *Science*. 120:684-686 (1954).

Freeman, N.G., A.M. Hale and M.B. Danard. "A Modified Sigma Equations Approach to the Numerical Modeling of Great Lakes Hydrodynamics," 77:1050-1060 (1972).

Freeman, N.G., and T.S. Murty, "A Study of a Storm Surge on Lake Huron," Proc. of the 15th Conf. of Great Lakes Res., Int. Assoc. Great Lakes Res. (1972), 565-582.

Freeman, N.G., P.F. Hamblin and T.S. Murty. "Helmholtz Resonance in Harbours of the Great Lakes," Proc. of the 17th Conf. of Great Lakes Res., Int. Assoc. Great Lakes Res. (1974), pp. 399-411.

Gallagher, R.H., J.A. Liggett and S.K.T. Chan. "Finite Element Shallow Lake Circulation," *J. Hydraul. Div. ASCE*. 99:1083-1096.

Gedney, R.T., and W. Lick. "Wind-Driven Currents in Lake Erie," *J. Geophys. Res*. 77:2714-2723 (1972).

Gedney, R.T., W. Lick and F.B. Molls. "A Simplified Stratified Lake Model for Determining Effects of Wind Variation and Eddy Diffusivity," Proc. of the 16th Conf. of Great Lakes Res., Int. Assoc. Great Lakes Res. (1973), pp.710-722.

Gorham, E., and F.M. Boyce. "The Influence of Lake Surface Area and Maximum Depth upon Thermal Stratification and the Depth of the Summer Thermocline," *J. Great Lakes Res*. 15:233-245 (1989).

Hamblin, P.F. "Hydraulic and Wind-Induced Circulation in a Model of a Great Lake," Proc. of the 12th Conf. of Great Lakes Res., University of Michigan, Great Lakes Res. Div. (1969).

Hamblin, P.F. "Some Free Oscillations of a Rotating Natural Basin," Ph.D. thesis, Univ. of Washington, Department of Oceanography (1972).

Hamblin, P.F. "A Theory of Short Period Tides in a Rotating Basin," *Phil. Trans. R. Soc. Lond*. A281:97-111 (1976).

Hamblin, P.F. "Great Lakes Storm Surge of April 6, 1979," *J. Great Lakes Res*. 5:312-315 (1979).

Hamblin, P.F. "Meteorological Forcing and Water Level Fluctuations on Lake Erie," *J. Great Lakes Res*. 5:312-315 (1987).

Harleman, D.R.F., R.M. Bunker and J.B. Hall. "Circulation and Thermocline Development in a Rotating Lake Model," Proc. of the 7th Conf. of Great Lakes Res., University of Michigan, Great Lakes Res. Div. Publ. No. 11 (1964), pp. 340-356.

Harrington, M.W. "Currents of the Great Lakes as Deduced from the Movements of Bottle Papers during the Seasons of 1892 and 1893," U.S. Weather Bureau, Washington, D.C. (1894).

Harris, R.A. "Manual of Tides," U.S. Coast and Geol. Surv. Rep., Washington, D.C. (1907), p. 483.

Harris, D.L. "Wind Tide and Seiches in the Great Lakes," Fourth Proc. of the Coastal Eng. Conf., Chicago, IL (1953), pp.25-51.

Harris, D.L., and A. Angelo. "A Regression Model for Storm Surge Prediction," *Mon. Weather Rev*. 91:710-726 (1963).

Hayford, J.F. "Effects of Winds and Barometric Pressures on the Great Lakes," Carnegie Institute of Washington (1922).

Heaps, N.S., C.H. Mortimer and E.J. Fee. "Numerical Models and Observations of Water Motion in Green Bay, Lake Michigan," *Phil. Trans. R. Soc. London.* A306:371-398 (1982).

Henry, A.J. "Wind Velocity and Fluctuations of Water Level on Lake Erie," U.S. Weather Bureau Bulletin J (1922).

Howell, J.A., K.M. Kiser and R.R. Rumer. "Circulation Patterns and a Predictive Model for Pollutant Distribution in Lake Erie," Proc. of the 13th Conf. of Great Lakes Res., Int. Assoc. of Great Lakes Res. (1970), pp. 434-443.

Huang, J.C.K. "The Thermal Current in Lake Michigan," *J. Phys. Oceanogr.* 1:105-122 (1971).

Huang, J.C.K., and J.H. Saylor, "Vorticity Waves in a Shallow Basin," *Dyn. Atmos. Oceans.* 6:177-196 (1982).

Hunt, I.A. "Winds, Wind Set-Ups, and Seiches on Lake Erie," U.S. Army Eng. Dist. Detroit, MI. (1959).

Irish, S.M., and G.W. Platzman. "An investigation of the meteorological conditions associated with extreme wind tides on Lake Erie," *Mon. Weather Rev.* 90:39-47 (1962).

Ivey, G.N., and F.M. Boyce. "Entrainment by Bottom Currents in Lake Erie," *Limnol. Oceanogr.* 27:1029-1038 (1982).

Ivey, G.N., and J.C. Patterson. "A Model of the Vertical Mixing in Lake Erie in Summer," *Limnol. Oceanogr.* 29:553-563 (1984).

Keulegan, G.H. "Wind Tides in Small Closed Channels," *J. Res. Natl. Bur. Stand.* 46:358-381 (1951).

Keulegan, G.H. "Hydrodynamic Effects of Gales on Lake Erie," *J. Res. Natl. Bur. Stand.* 50:99-110 (1953).

Kizlauskas, A.G., and P.L. Katz. "A Two-Layer Finite Difference Model for Flows in Thermally Stratified Lake Michigan," Proc. of the 16th Conf. of Great Lakes Res., Int. Assoc. Great Lakes Res. (1973) pp. 743-753.

Lam, D.C.L. "Simulation of Water Circulations and Chloride Transports in Lake Superior for Summer 1973," *J. Great Lakes Res.* 4:343-349 (1978).

Lam, D.C.L., and T.J. Simons. "Numerical Computations of Advective and Diffusive Transports of Chloride in Lake Erie during 1970," *J. Fish. Res. Board Can.* 33:537-549 (1976).

Lam, D.C.L., W.M. Schertzer and A.S. Fraser. "Simulation of Lake Erie Water Quality Responses to Loading and Weather Variations," *Environ. Can. Sci. Ser.* 134 (1983).

Lam, D.C.L., and W.M. Schertzer. "Lake Erie Thermocline Model Results: Comparison with 1967-1982 Data and Relation to Anoxic Occurrences," *J. Great Lakes Res.* 13:757-769 (1987).

Lam, D.C.L., W.M. Schertzer and A.S. Fraser. "Oxygen Depletion in Lake Erie: Modeling the Physical, Chemical, and Biological Interactions, 1972 and 1979," *J. Great Lakes Res.* 13(4):770-781 (1987a).

Lam, D.C.L., W.M. Schertzer and A.S. Fraser. "A Post Audit Analysis of the NWRI Nine-Box Water Quality Model for Lake Erie," *J. Great Lakes Res.* 13(4):782-800 (1987b).

Lee, K.K., and J.A. Liggett. "Computation for Circulation in Stratified Lakes," *J. Hydraul. Div. ASCE.* 96:2089-2115 (1970).

Lee, N.Y., W. Lick and S.W. Kang. "The Entrainment and Deposition of Fine-Grained Sediments in Lake Erie," *J. Great Lakes Res.* 7:224-233 (1981).

Lesht, B.M., and N. Hawley. "Near Bottom Currents and Suspended Sediment Concentration in Southeastern Lake Michigan," *J. Great Lakes Res.* 13:375-383 (1987).

Li, C.Y., K.M. Kiser and R.R. Rumer. "Physical Model Study of Circulation Patterns in Lake Ontario," *Limnol. Oceanogr.* 20:323-337 (1975).

Lien, S.L., and J.A. Hoopes. "Wind-Driven Steady Flows in Lake Superior," *Limnol. Oceanogr.* 23:91-103 (1978).

Liggett, J.A. "Unsteady Circulation in Shallow, Homogeneous Lakes," *J. Hydraul. Div. ASCE.* 95:1273-1288 (1969).

Liggett, J.A. "Cell Method for Computing Lake Circulation," *J. Hydraul. Div. ASCE.* 96:725-743 (1970).

Liggett, J.A., and C. Hadjitheodorou. "Circulation in Shallow Homogeneous Lakes," *J. Hydraul. Div. ASCE.* 95:609-620 (1969).

Liggett, J.A., and K.K. Lee. "Properties of Circulation in Stratified Lakes," *J. Hydraul. Div. ASCE.* 97:15-29 (1971).

Liu, P.C., D.J. Schwab and J.R. Bennett. "Comparison of a Two-Dimensional Wave Prediction Model with Synoptic Measurements in Lake Michigan," *J. Phys. Oceanogr.* 14:1514-1518 (1984).

Marmorino, G.O. "Inertial Currents in Lake Ontario, Winter 1972-73 (IFYGL)," *J. Phys. Oceanogr.* 8:1104-1120 (1978).

McCormick, M.J., and G.A. Meadows. "An Intercomparison of Four Mixed Layer Models in a Shallow Inland Sea," *J. Geophys. Res.* 93:6774-6788 (1988).

Millar, F.G. "Surface Temperatures of the Great Lakes," *J. Fish. Res. Board Can.* 9:329-376 (1952).

Mortimer, C.H. "Frontiers in Physical Limnology with Particular Reference to Long Waves in Rotating Basins," Proc. of the 6th Conf. of Great Lakes Res., University of Michigan Great Lakes Res. Div., Publ. No. 10 (1963), pp. 9-42.

Mortimer, C.H. "Spectra of Long Surface Waves and Tides in Lake Michigan and Green Bay, Wisconsin," Proc. of the 8th Conf. of Great Lakes Res., University of Michigan Great Lakes Res. Div., Publ. No. 13 (1965), pp. 304-325.

Mortimer, C.H. "Internal Waves and Associated Currents Observed in Lake Michigan during the Summer of 1963," University of Wisconsin-Milwaukee, Center of Great Lakes Studies, Special Report No. 1 (1968).

Mortimer, C.H. "Lake Hydrodynamics," *Mitt. Int. Ver. Theor. Agnew. Limnol.* 20:124-197 (1974).

Mortimer, C.H. "Measurements and Models in Physical Limnology," in *Hydrodynamics of Lakes: CISM Lectures*, K. Hutter, Ed. (New York: Springer Verlag, 1984), pp. 287-322.

Mortimer, C.H., and E.J. Fee. "Free Oscillations and Tides of Lakes Michigan and Superior," *Phil. Trans. R. Soc. Lond.* A281:1-61 (1976).

Murthy, C.R., T.J. Simons and D.C.L. Lam. "Simulation of Pollutant Transport in Homogeneous Coastal Zones with Application to Lake Ontario," *J. Geophys. Res.* 91:9771-9779 (1986).

Murty, T.S., and D.B. Rao. "Wind-Generated Circulations in Lakes Erie, Huron, Michigan, and Superior," Proc. of the 13th Conf. of Great Lakes Res., Int. Assoc. Great Lakes Res. (1970), pp. 927-941.

Murty, T.S., and R.J. Polavarapu. "Reconstruction of Some of the Early Storm Surges on the Great Lakes," *J. Great Lakes Res.* 1:116-129 (1975).

Paul, J.F., and W.J. Lick. "A Numerical Model for Thermal Plumes and River Discharges," Proc. 17th Conf. Great Lakes Res. Int. Assoc. Great Lakes Res. (1974), pp. 445-455.

Pickett, R.L. "Observed and Predicted Great Lakes Winter Circulations," *J. Phys. Oceanogr.* 10:1140-1145 (1980).

Pickett, R.L., and D.A. Dossett. "Mirex and the Circulation of Lake Ontario," *J. Phys. Oceanogr.* 9:441-445 (1979).

Platzman, G.W. "A Numerical Computation of the Surge of 26 June 1954 on Lake Michigan," *Geophysica.* 6:407-438 (1958).

Platzman, G.W. "The Dynamical Prediction of Wind Tides on Lake Erie," *Meteorol. Monogr.* 4(26):44 (1963).

Platzman, G.W. "The Prediction of Surges in the Southern Basin of Lake Michigan. Part 1. The Dynamical Basis for Prediction," *Mon. Weather Rev.* 93:275-281 (1965).

Platzman, G.W. "The Daily Variation of Water Level on Lake Erie," *J. Geophys. Res.* 71:2471-2483 (1966).

Platzman, G.W. "Two-Dimensional Free Oscillations on Natural Basins," *J. Phys. Oceanogr.* 2:117-138 (1972).

Platzman, G.W., and D.B. Rao. "The Free Oscillations of Lake Erie," in *Studies on Oceanography*, K. Yoshida, Ed. (Tokyo: University of Tokyo Press, 1964a), pp. 359-382.

Platzman, G.W., and D.B. Rao. "Spectra of Lake Erie Water Levels," *J. Geophys. Res.* 69:2525-2535 (1964b).

Pore, N.A. "Automated Wave Forecasting for the Great Lakes," *Mon. Weather Rev.* 107:1275-1286 (1979).

Quinn, F.A. "Hydrologic Response Model of the North American Great Lakes," *J. Hydrol.* 37:295-307 (1978).

Rao, D.B., and T.S. Murty. "Calculation of Wind-Driven Circulations in Lake Ontario," *Arch. Meteorol. Geophys. Bioklimatol.* A19:195-210 (1970).

Rao, D.B., and D.J. Schwab. "Two-Dimensional Normal Modes in Arbitrary Enclosed Basins on a Rotating Earth: Application to Lakes Ontario and Superior," *Phil. Trans. R. Soc. London.* 281A:63-96 (1976).

Rao, D.B., C.H. Mortimer and D.J. Schwab. "Surface Normal Modes of Lake Michigan: Calculations Compared with Spectra of Observed Water Level Fluctuations," *J. Phys. Oceanogr.* 6:575-588 (1976).

Richardson, W.S., and N.A. Pore. "A Lake Erie Storm Surge Forecasting Technique," ESSA Technical Memo. WBTM TDL 24, NTIS PB-187778 (1969).

Robertson, A. and D. Scavia. "The Examination of Ecosystem Properties of Lake Ontario through the use of an ecological model," in *Perspectives on Lake Ecosystem Modeling*, D. Scavia and A. Robertson, Eds. (Ann Arbor, MI: Ann Arbor Science Publishers, 1979), pp. 281-292.

Rockwell, D.C. "Theoretical Free Oscillations of the Great Lakes," Proc. 9th Conf. Great Lakes Res. Univ. Mich., Great Lakes Res. Div., Publ. No. 15 (1966), pp. 352-368.

Rodgers, G.K. "The Thermal Bar in the Laurentian Great Lakes," Proc. of the 8th Conf. of Great Lakes Res., Univ. Mich. Great Lakes Res. Div., Publ. No. 13 (1965), pp. 358-363.

Rodgers, G.K. "The Thermal Bar in Lake Ontario, Spring 1965 and Winter 1965-66," Proc. of the 9th Conf. of Great Lakes Res., Univ. Mich., Great Lakes Res. Div., Publ. No. 15 (1966), pp. 369-374.

Rumer, R.R., and L. Robson. "Circulation Studies in a Rotating Model of Lake Erie," Proc. of the 11th Conf. of Great Lakes Res., Int. Assoc. Great Lakes Res. (1968), pp. 487-495.

Saylor, J.H., J.C.K. Huang, and R.O. Reid. "Vortex Modes in Southern Lake Michigan," *J. Phys. Oceanogr.* 10:1814-1823 (1980).

Scavia, D. "Examination of Phosphorous Cycling and Control of Phytoplankton Dynamics in Lake Ontario with an Ecological Model," *J. Fish. Res. Board Can.* 36:1336-1346 (1979).

Scavia, D. "An Ecological Model of Lake Ontario," *Ecol. Modeling* 8:49-78 (1980).

Scavia, D., and J.R. Bennett. "Spring Transition Period in Lake Ontario - A Numerical Study of the Causes of Large Biological and Chemical Gradients," *Can. J. Fish. Aquat. Sci.* 37:823-833 (1980).

Schertzer, W.M., J.H. Saylor, F.M. Boyce, D.G. Robertson and F. Rosa. "Seasonal Thermal Cycle of Lake Erie," *J. Great Lakes Res.* 13:468-486 (1987).

Schwab, D.J. "Internal Free Oscillations in Lake Ontario," *Limnol. Oceanogr.* 22:700-708 (1977).

Schwab, D.J. "Simulation and Forecasting of Lake Erie Storm Surges," *Mon. Weather Rev.* 106:1476-1487 (1978).

Schwab, D.J. "An Inverse Method for Determining Wind Stress from Water-Level Fluctuations," *Dyn. Atmos. Oceans.* 6:251-278 (1982).

Schwab, D.J. "Numerical Simulation of Low-Frequency Current Fluctuations in Lake Michigan," *J. Phys. Oceanogr.* 13:2213-2224 (1983).

Schwab, D.J., and D.B. Rao. "Gravitational Oscillations of Lake Huron, Saginaw Bay, Georgian Bay, and the North Channel," *J. Geophys. Res.* 82:2105-2116 (1977).

Schwab, D.J., J.R. Bennett and E.W. Lynn. "'Pathfinder' — A Trajectory Prediction System for the Great Lakes," NOAA Technical Memo ERL GLERL-53 (1984).

Schwab, D.J., J.R. Bennett, P.C. Liu and M.A. Donelan. "Application of a Simple Numerical Wave Prediction Model to Lake Erie," *J. Geophys. Res.* 89:3586-3592 (1984).

Schwab, D.J., and J.R. Bennett. "Lagrangian Comparison of Objectively Analyzed and Dynamically Modelled Circulation Patterns in Lake Erie," *J. Great Lakes Res.* 13:515-529 (1987).

Schwab, D.J., A.H. Clites, C.R. Murthy, J.E. Sandall, L.A. Meadows and G.A. Meadows. "The Effect of Wind on Transport and Circulation in Lake St. Clair," *J. Geophys. Res.* 94:4947-4958 (1989).

Scott, J.T. and L. Lansing. "Gradient Circulation in Eastern Lake Ontario," Proc. of the 10th Conf. of Great Lakes Res., University of Michigan, Great Lakes Res. Div. (1967), pp. 322-336.

Sheng, Y.P., W.J. Lick, R.T. Gedney and F.B. Molls. "Numerical Computation of Three-Dimensional Circulation in Lake Erie: A Comparison of a Free Surface Model and a Rigid Lid Model," *J. Phys. Oceanogr.* 8:713-727 (1978).

Sheng, Y.P. and W. Lick. "The Transport and Resuspension of Sediments in a Shallow Lake," *J. Geophys. Res.* 84:1809-1825 (1979).

Simons, T.J. "Development of Numerical Models of Lake Ontario," Proc. of the 14th Conf. of Great Lakes Res., Int. Assoc. Great Lakes Res. (1971), pp. 654-669.

Simons, T.J. "Development of Numerical Models of Lake Ontario: Part 2," Proc. of the 15th Conf. of Great Lakes Res., Int. Assoc. Great Lakes Res. (1972), pp. 655-672.

Simons, T.J. "Development of Three-Dimensional Numerical Models of the Great Lakes," *Can. Inland Waters Branch. Sci. Ser.* 12:26 (1973).

Simons, T.J. "Verification of Numerical Models of Lake Ontario: I. Circulation in Spring and Early Summer," *J. Phys. Oceanogr.* 4:507-523 (1974).

Simons, T.J. "Verification of Numerical Models of Lake Ontario: II. Stratified Circulations and Temperature Changes," *J. Phys. Oceanogr.* 5:98-110 (1975).

Simons, T.J. "Continuous Dynamical Computations of Water Transports in Lake Erie for 1970," *J. Fish. Res. Board Can.* 33:371-384 (1976a).

Simons, T.J. "Verification of Numerical Models of Lake Ontario. III. Long-Term Heat Transports, *J. Phys. Oceanogr.* 6:372-378 (1976b).

Simons, T.J. "Analysis and Simulation of Spatial Variations of Physical and Biochemical Processes in Lake Ontario," *J. Great Lakes Res.* 2:215-233 (1976c).

Simons, T.J. "On the Joint Effect of Baroclinicity and Topography," *J. Phys. Oceanogr.* 9:1238-1287 (1979).

Simons, T.J. "Circulation Models of Lakes and Inland Seas," *Can. Bull. Fish. Aquat. Sci.* 203:146 (1980).

Simons, T.J. "Resonant Topographic Response of Nearshore Currents to Wind Forcing," *J. Phys. Oceanogr.* 13:512-523 (1983).

Simons, T.J. "Topographic Response of Nearshore Currents to Wind" An Empirical Model," *J. Phys. Oceanogr.* 14:1393-1398 (1984).

Simons, T.J. "Reliability of Circulation Models," *J. Phys. Oceanogr.* 15:1191-1204 (1985).

Simons, T.J. "The Mean Circulation of Unstratified Water Bodies Driven by Nonlinear Topographic Wave Interaction," *J. Phys. Oceanogr.* 16:1138-1142 (1986).

Simons, T.J., G.S. Beal, K. Beal, A.H. El-Shaarawi and T.S. Murty. "Operational Model for Predicting the Movement of Oil in Canadian Navigable Waters," Manuscript Report Series, No. 37, Marine Science Directorate, Department of the Environment, Ottawa, Ontario, Canada (1975).

Simons, T.J., and D.C.L. Lam. "Some Limitations of Water Quality Models for Large Lakes: A Case Study for Lake Ontario," *Water Resour. Res.* 16:105-116 (1980).

Simons, T.J., and W.M. Schertzer. "Modeling Wind-Induced Setup in Lake St. Clair," *J. Great Lakes Res.* 15:452-464 (1989).

Snodgrass, W.J. "Analysis of Models and Measurements for Sediment Oxygen Demand in Lake Erie," *J. Great Lakes Res.* 13(4):738-756 (1987).

Thomann, R.V., R.P. Winfield and D.Z. Szumski. "Estimated Responses of Lake Ontario Phytoplankton Biomass to Varying Nutrient Levels," *J. Great Lakes Res.* 3(1-2):123-131 (1977).

Thomann, R.V., D.M. DiToro, D. Scavia and A. Robertson. "Ecosystem and Water Quality Modeling. IFYGL - The International Field Year for the Great Lakes," E.J. Aubert and T.L. Richards Eds. (Ann Arbor, MI: NOAA Great Lakes Environ. Res. Lab., 1981), pp. 353-366.

Thomas, J.H. "A Theory of Steady Wind-Driven Currents in Shallow Water with Variable Eddy Viscosity," *J. Phys. Oceanogr.* 5:136-142 (1975).

Verber, J.L. "Detection of Rotary Currents and Internal Waves in Lake Michigan," Proc. of the 7th Conf. of Great Lakes Res., Univ. Mich. Great Lakes Res. Div. (1964), pp. 382-389.

Verber, J.L. "Inertial Currents in the Great Lakes," Proc. of the 9th Conf. of Great Lakes Res., Univ. Mich., Great Lakes Res. Div., Publ. No. 15 (1966), pp. 375-379.

Witten, A.J. and J.H. Thomas. "Steady Wind-Driven Currents in a Large Lake with Depth-Dependent Eddy Viscosity," *J. Phys. Oceanogr.* 6:85-92 (1976).

3 SEDIMENT EXCHANGE PROCESS PARAMETERIZATION

INTRODUCTION

Parameterizing the relationship between sediment transport and the water column distribution of certain organic and inorganic compounds is an essential task required for predicting the fate and transport of these often hazardous substances. With the range of freshwater particle sizes typically being from 1 to 200 μm, the force imbalance giving rise to sediment motions can result from an almost bewildering variety of fluid mechanical, material and molecular mechanisms which are often extremely difficult if not impossible to measure in the field or credibly parameterize. Yet the press for such parameterization continues as the use of surface water models for assessing the impact of these chemicals, while not a simple task, is almost standard practice.

The ephemeral, selective and differential nature of the sediment response to fluid forces makes the parameterization task difficult as the predictive expression must cover such a wide variety of possible circumstances. However, such generality is rarely obtained as the bases for most parameterizations are theoretical solutions or Buckingham Pi groupings which are obtained from arguments limited to circumstances when one or two of the many forces are dominant. Consequently, a wide variety of empirical representations for aspects of the above sediment transport modeling exist. Four classes of modeling activity are necessary in order to predict sediment/chemical distributions; the first being predictions of basic circulation features of the water body as forced by wind fields and tributary inflows; the second being the prediction of basin wide sediment distributions in response to the hydrodynamic field for each grain size class; the third comprising models of how the particles interact with the chemical compounds (i.e., the source/sink terms); and the fourth being bed and exchange process characterization.

This chapter reviews the fourth submodel category of bed/exchange processes. The viewpoint adopted here will not be to simply write another hydrodynamics and transport review article as a number of them exist already, i.e., Grant

ISBN 0-87371-511-X
© 1992 by Lewis Publishers

and Madsen (1986), Bedford and Abdelrhman (1987), Gust (1984), Bechteler (1986), along with earlier works by McCave (1976), Bowden (1978), Nihoul (1977), and Nowell (1983). Rather, with the emphasis of this book speaking to water quality analysts, the objectives of this chapter are (1) to give insight into the physical complexity of the problem, (2) to summarize the terms and definitions used to describe the processes and resulting sediment fluxes, (3) to review the methods for parameterizing these exchanges, (4) to summarize the limiting assumptions, and (5) to present, where available, field evidence to support or rebut various ideas. While some comments will be made about "bed models", the major emphasis here (as well as for regulatory concerns) is in quantifying "resuspension," a confusing term used quite frequently but often understood in as many ways as the numbers of people using it. The article will begin with a brief description of the physical setting for which the exchange processes are to be described before setting forth definitions.

PHYSICAL SETTING COMPLEXITY

Bottom exchange processes are exceptionally complex because of the extremely wide range of length and time scales of the effective physical processes. Ranging from interparticle tractive forces operating on scales of microns to large scale motions of the entire water body operating at scales of kilometers, these processes interact, often nonlinearly, to create a net transfer of sediment to or from the bottom. Figure 1 portrays a space-time diagram of the various short- and long-term fluid mechanical processes in a typical body of water as scaled, for illustrative purposes, to the Great Lakes. Two clusters of processes appear; the first being larger scale processes dominated by wave-like motions including tides, Poincairé and Kelvin waves, seiches, etc. The second cluster consists of smaller scale processes such as turbulence and wind waves. In attempting to place these data in context, a useful approach has been to explore how energy is transmitted between these processes, as it is energy that must be "delivered" to the bottom in order to do the work involved in the exchange process.

A variety of approaches exist with which to characterize energy transformations, but a combination of energy cascade (Tennekes and Lumley 1972) and nonlinear system dynamics (Baker and Gollub 1990, Thompson and Stewart 1986) descriptions seems appropriate. Figure 2 from Woods (1977), (as slightly augmented by Bedford 1991b), gives a schematic of the energy flow. Energy is input to the system at both the large an small scale process clusters.

The large scale inputs are in the form of atmospheric low pressure and wind stress frontal motions while the small scale inputs are in the form of wind waves. The energy is then seen to basically proceed from the large scale to the small scale where it is dissipated by friction and molecular scale processes. The energy transfer mechanisms are extremely complex but can be separated into four zones of energy conversion as defined by the Rossby (R_o), Richardson (R_i) and Reynolds (R_e) numbers. While a fairly traditional energetics description, some additional

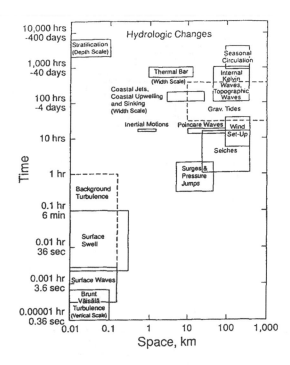

Figure 1.   **Block space-time diagram of water column fluid mechanical processes in the Great Lakes.**

points are noted. First, the energy conversion scenarios are extremely complex, particularly in the case of stratified flows, and consist of much more than inertial interactions. The review articles by Gregg (1987) and Hopfinger (1987) give a very clear summary of this complexity. Second, well "within" each of the energy conversion domains equilibrium or attractor states are the norm where the effective processes are few in number and conversion and transfer rates are predictable and theoretically describable. Third, at the boundaries between these regimes the overlapping of the various processes from each requires nonequilibrium dynamics descriptions that are often characterized by intermittencies. Fourth, this simple energetics picture must be augmented to account for additional intermittencies or dynamic forcings imposed by the weather/wind systems. In the case of the Great Lakes (Bedford and Abdelrhman 1987), storms often arrive more frequently than it takes many of the large scale energy conversion processes to come to equilibrium from a prior disturbance. While theoretical arguments (not yet available in the literature) will show whether all the intermittencies in these systems are self-similar and therefore predictable, it surely can be concluded that the nature of the flow affecting the bottom is dynamic, nonstationary and intermittent over all scales. Equilibrium, which is ubiquitous in theoretical formulations, is rarely observed in the field.

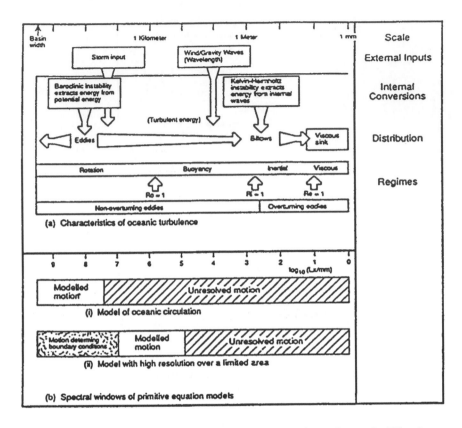

Figure 2.   Energetics processes and transformation schematic (Woods 1977 as augmented in Bedford 1991b).

The separation between the two clusters noted in Figure 1 is not an artifact; it occurs naturally as the separation (Figure 2) between processes which are three dimensional in the sense of complete overturning and which are quick to adjust to new equilibrium ($R_e > 1$, $R_i < 1$) and the larger scale processes with two-dimensional variability ($R_i > 1$, $R_o \gtrsim 1$) which are slow to evolve and adjust to a new equilibrium. It is tempting to exploit this separation as a means of developing procedures for parameterizing the exchange processes.

INFORMATION NEEDS

As noted in the introduction, it is this author's opinion that there is a bewildering variety of terminology which has arisen from different disciplines attempting to describe the same feature or process. This terminology very much confuses the communication of ideas across disciplines such as between geologists, chemists, and physical limnologists and between user classes such as scientists, water quality managers, administrators, and politicians. Therefore, the first information need is a standardized terminology.

Terminology-An Equation Based Approach

In order to understand the basis for the resuspension definition and subsequent parameterization it is necessary to summarize equations which portray the fluid mechanical processes contributing to the observed sediment concentration variation in time at a fixed point. It is assumed that the flow is by and large turbulent although nearbottom viscous sublayers have been observed in the ocean (Chriss and Caldwell 1984). The bottom is assumed to be rough with four possible regimes defined: (1) a horizontal bed where roughness is due to sand grains, (2) weak bedforms, i.e., ripples on a horizontal bed or roughness elements where roughness density (Nowell and Church 1979) gives rise to skimming, wake interaction and isolated roughness flows, (3) a macro bedform regime, i.e., dunes whose scale is large compared to the boundary layer sampling scale, and (4) a washed out bed form regime.

A coordinate system is defined in the plane of the bottom with the z-coordinate being positive away from the bottom and parallel to gravity. Coordinate systems placed on weakly sloping bottoms introduce small vertical coordinate departures from gravity whose contribution to resolved horizontal effects on concentration are small but whose effect on correlations such as turbulent sediment flux are large (Grant and Madsen 1986, Bedford 1991a) if not adequately accounted for in the analyses (Bedford et al. 1987).

Based upon the reviews in Lumley (1978) and Van Rijn (1986), the basic equation describing the conservation of sediment mass at a point may be derived from four points of view; in either total concentration or sediment size class forms for sparse or dense particle count mixtures. In the first class of formulations, the total mass or volume concentration is the dependent variable and the settling velocity is the ensemble average for the grain size mixture which therefore must vary in time or space in accordance with the size distribution. In the second formulation a mass conservation equation is written for each size class and solved assuming a constant settling velocity for each size class. As discussed in Lumley (1978), it is extremely rare that the dense particle formulation is required, therefore, for this discussion, the standard sparse particle mass or volume concentration formulation is used.

A shortcut often used in the resulting equation is to assume a streamline following coordinate system, which means that the two horizontal velocities are vector summed into an average horizontal velocity which allows a reduction of the equation to two dimensions (one vertical, one horizontal). Implied here is that the direction of the horizontal velocity is known or must be measured. In the field the direction of the average horizontal velocity is changing over time and is often in a direction different than the wind wave trains.

Finally, it is not possible to adequately incorporate the effect of all the relevant scales of variability with one general formulation. One approach then is to construct theories that reflect only those dominant influences leading to the various equilibrium states mentioned previously. This approach leads to parameterizations valid near the local equilibrium states. The second analysis approach is to average or filter the equations and remove small scale fluctuations from the basic equations. Most resuspension parameterizations now in use by models are derived from the first approach; while most theoretical treatments, especially those derived from a fluid mechanical perspective, are developed from the second approach. With very careful analysis and data preparation, both approaches can converge to the same resulting description.

Procedures for averaging the basic equations are well documented and principally based on uniform weight function averaging procedures originally used by O. Reynolds in 1882. This procedure essentially limits the equations to variables that are steady in the mean over the averaging period, T, and leaves fluctuations with stationary statistics (Monin and Yaglom 1975). Therefore, use of Reynold's average in the field suggests that T can be no larger than what is required for stationary fluctuation statistics, typically 10 to 15 minutes (Bedford et al., 1987). By use of Reynold's rules of averaging as applied over T, the temporal average concentration equation as a function of a time (t) and space (x_i) is

$$\frac{D\bar{c}}{Dt} = \frac{\partial \bar{c}}{\partial t} + \frac{\partial \bar{N}_i}{\partial x_i} = S \tag{1}$$

where the overbar indicates the time average variable defined as

$$\bar{\alpha} = \frac{1}{T} \int_t^{t+T} \alpha \, dt \tag{2}$$

The subscript denotes the three components of the spatial coordinates i = 1,2,3 corresponding to the x,y,z directions, respectively. S is the source/sink term denoting the increase or decrease of $\bar{c}$ in the volume by means other than velocity induced flux. The function N_i is the flux of concentration with dimensions of mass per area perpendicular to the direction i per time, ($ML^{-2}t^{-1}$). With the fluid velocity vector v_i having components in the x($\rightarrow$ u), y($\rightarrow$ v), and z($\rightarrow$ w) directions, respectively, the flux is further defined as

$$\bar{N}_i = \bar{u}_i \bar{c} + \overline{u_i' c'} - w_s \bar{c} \cos \theta_i - D \frac{\partial \bar{c}}{\partial x_i} \tag{3}$$

Therefore flux is a vector quantity with three components. Should the local coordinates correspond to a flat bottom (z→ gravity), then $\cos\theta_1 = \cos\theta_2 = 0$ and $\cos\theta_3 = 1$. w_s is the ensemble or aggregate setting velocity for the mixture, which adjusts as the grain size distribution changes. The prime terms are fluctuations relative to the local average value, and while $\overline{\alpha'} = 0$ the time average of the product of fluctuations is not unless the variables are not at all correlated. The parameter D is the molecular diffusion coefficient for particles in water which is essentially a property of the fluid (Vanoni 1975). Its value is based upon Brownian motion arguments and is usually so small in comparison to the other terms that the term is typically dropped.

The definition of flux terms proceeds further. If gravity coincides with z, then the horizontal flux is written in a streamline coordinate x as

$$\overline{N_x} = \overline{uc} + \overline{u'c'} - D \frac{\partial \overline{c}}{\partial x} \tag{4}$$

while the vertical flux (+ away from bottom) is written as

$$\overline{N_z} = \overline{wc} + \overline{w'c'} - w_s \overline{c} - D \frac{\partial \overline{c}}{\partial z} \tag{5}$$

A special case exists for vertical flux at the bottom $z = 0$. Here, the mean vertical velocity is zero and the vertical flux at the bottom becomes

$$\overline{N_z}(z = 0) = \overline{N_{zo}} = \overline{w'c'} - w_s \overline{c} - D \frac{\partial \overline{c}}{\partial z} \tag{6}$$

All of the terms comprising the fluxes in equations (4) to (6) have specific names associated with them and Table 1 summarizes these labels. It should be noted that all these fluxes, except the settling (Term e) and entrainment (Term g) fluxes, can in principle have either plus or minus signs; settling as defined can only have a component towards the bed; while turbulent flux at the bottom is by weight of field evidence, almost always away from the bottom due to asymmetric turbulence fields whose scale increases with distance away from the wall (Hinze 1975).

If, instead of the total concentration, $\overline{c}$, and aggregate settling velocity, w_s, one formulates the equations from a grain-size class point of view, then a conservation equation for each class ℓ is written with $\overline{c}_\ell$ as the dependent variable, along with a constant settling velocity $w_{s\ell}$ for each class l. The definitions in Table 1 still apply for each class. The total flux $\overline{N_i}$ would then be comprised of the mass fluxes of the individual components, i.e.,

Table 1.   Definition of Flux Terms

Definition	Eq. Form	Eq. No.
a. Horizontal flux also suspended load	$\overline{N}_x or (q_{ss})$	3, 4
b. Vertical flux	$\overline{N}_z$	3, 5, 6
c. Advective flux		
i). Horizontal	$\overline{u}\,\overline{c}$	3, 4
ii). Vertical	$\overline{w}\,\overline{c}$	3, 5, 6
d. Turbulent flux		
i). Horizontal	$\overline{u'c'}$	3, 4
ii). Vertical	$\overline{w'c'}$	3, 5, 6
e. Settling flux	$-w_s \overline{c}$	3, 5, 6
f. Molecular flux		
i). Horizontal	$-D \partial \overline{c} / \partial x$	3, 4
ii). Vertical	$-D \partial \overline{c} / \partial z$	3, 5, 6
g. Entrainment flux	$\overline{w'c'}(z = 0)$	6

$$\overline{N}_i = \sum_{\ell=1}^{M} \overline{N}_{i\ell} \tag{7}$$

where M equals the total number of grain size classes. In most large scale modeling applications three classes are typically used (e.g., Ziegler and Lick 1986, Sheng 1983, Lee and Bedford 1987).

Based upon the direction of the total (or class size) flux, three additional pieces of flux terminology are derived as in Table 2. The definitions in Table 2 differ from those in Table 1 in that Table 1 defines the process names while Table 2 defines the net aggregate of the processes. The net bottom flux is once again distinguished from the general water column flux because net gain or loss of mass through the control volume boundary at the bottom is clearly and unambiguously labelled and separated from water column processes.

It is quite easy to confuse these labels, not only because there are so many, but because the two formulation classes are often "blended" when creating parameterizations for bed exchange processes that are consistent with both intuition and theory. As an example, from the mixture formulation approach, fluid mechanical conditions might result in a mixture vertical flux that is in

Table 2.   Derived Flux Terminology

	Label	Definition
Water column	Depositional flux	$N_z < 0$
	Equilibrium flux	$N_z = 0$
	Resuspension flux	$N_z > 0$
Bottom exchange	Sedimentation flux	$N_{zo} < 0$
	Equlibrium flux	$N_{zo} = 0$
	Erosion flux	$N_{zo} > 0$

equilibrium, $N_{zo} = 0$, but be comprised of equal sedimenting sand fraction and erosional clay fraction fluxes. While this example speaks more to formulation inadequacies, it nevertheless points out the necessity of clear terminology. Since fluxes are very difficult and expensive to measure, a thorough understanding of which flux is being addressed will certainly minimize costs and resulting misunderstandings.

An additional terminology confusion derives from specifying "closure" for the turbulent correlation or flux term $\left(\overline{u'c'}, \overline{w'c'} \right)$. Since Prandtl, it has been customary to use the Boussinesq analogy (Hinze 1975) and set these parameters proportional to the mean concentration gradient in the respective direction; i.e.,

$$\overline{u'c'} = -E_x \frac{\partial \bar{c}}{\partial x}; \quad \overline{w'c'} = E_z \frac{\partial \bar{c}}{\partial z} \tag{8}$$

In Equation 8 E_x and E_z are each called the eddy diffusivity (Hinze 1975) or the turbulent diffusion coefficient (Vanoni 1974) and unlike the molecular diffusion coefficient are not properties of the fluid but are a result of the turbulent characteristics of the flow. A review of these parameterizations is found in the papers by the American Society of Civil Engineers (ASCE 1988). These coefficients are *not* dispersion coefficients nor are the processes they represent called dispersion. Dispersion (ASCE 1988, Fischer et al. 1979) represents the effect of "closure" resulting from spatial averaging over one or more spatial domains in the flow field. For example (Fischer et al. 1979) longitudinal bulk dispersion coefficients in river pollutant transport models result from averaging the transport equation over the cross-sectional area of the river. Dispersion tensors arise from two-dimensional formulations based upon averaging over the local water depth (e.g., Holly and Usseglio-Polatera 1974). The mechanisms giving rise to the eddy diffusivity values have a direct bearing on the bottom exchange processes while dispersion processes only affect the representation of whole water column transport and not bottom exchange processes.

EXCHANGE PROCESS PARAMETERIZATIONS

From Bedford (1990a), there are three procedures by which parameterizations for exchange processes can be formulated: (1) dimensional analysis/empirical, (b) theoretical/mechanistic, and (c) structural. In the first, methods of Buckingham Pi and similarity analyses are used to derive empirical functions with specific regressions detailed from laboratory experiments; the second uses the theoretical equations governing sediment transport [e.g., Equations 1 to 6] to achieve comprehensive solutions to specific sets of conditions; and the third attempts to detail the structure of the data contributing to the observed variations in the fluid processes. The third approach, while offering scientific insight, has not contributed to predictive parameterizations. Because the predictive boundary layer theories are an important tool, method number two has been relentlessly pursued in parallel along with the more traditional laboratory oriented empirical procedures. Reconciling the results and methods of the first two approaches is a good deal more difficult than might first be imagined. Lack of consistency in adapting and comparing the two methods is a major source of the difficulty encountered in attempting to compare predictions made with the exchange process parameterizations to laboratory and field data.

This section summarizes the functional forms now available for parameterizing the sedimentation and erosional fluxes at the bed as extracted from the report by Lee and Bedford (1987) and the articles by Mehta et al. (1989) and those found in the conference proceedings edited by Mehta (1986) and Mehta and Hayter (1989). The following section summarizes their data input requirements and assesses the status of field, laboratory, and theoretical treatments for obtaining these input data.

Bottom exchange models are currently available for erosional fluxes and sedimentation fluxes and within each of these classes the distinction is further made between cohesive and noncohesive formulations. Additionally, erosion models are further refined as to whether the bottom consists of consolidated/compacted sediments or recently deposited unconsolidated sediments.

Mathematically, exchange processes are formally incorporated by specifying the boundary condition for equation (1) at $z = 0$. This boundary condition can take two possible forms.

$$\bar{c}(z = 0, x, t) = c * (x, t) \tag{9a}$$

$$\bar{N}_z(z = 0, x, t) = \bar{N}_{zo} = N * (x, t) \tag{9b}$$

In other words, either the flux or the concentration are specified at the bottom. In transport model practice, however, entrainment and sedimentation fluxes are parameterized as source (S_E) and sink (S_D) terms as in Equation 1, i.e.,

$$\frac{D\bar{c}}{Dt} = \frac{\partial \bar{c}}{\partial t} + \frac{\partial \bar{N}_i}{\partial x_i} = S_E - S_D \tag{10}$$

Essentially then, for the control volume bed area over a small time (~dt), this amounts to an incremental adjustment in concentration.

Erosion Flux Parameterization

Table 3 summarizes forms of the erosion flux parameterization, which in this author's opinion are the most frequently used. In many ways based upon the thin film convective transfer approaches developed by chemical engineers (Bennet and Meyers 1974), these formulations for the most part posit that the transfer of material to the water column results from a nonequilibrium force balance between the force applied by the fluid to the bottom and the restraining forces existing within the bed material. It is often assumed (see Vanoni 1975 for an excellent historical review) that the most fundamental measures of the fluid-bed force imbalance are the fluid and bed shear stress. Erosion flux then is based upon whether the local fluid shear stress at the bottom τ_b, is greater (erosion) or less (no erosion) than the material/particle critical shear stress of the bed mixture (Krone et al. 1977) required for erosion. Once the fluid shear exceeds the critical shear stress for erosion, the erosion rate linearly increases for consolidated bottom material and nonlinearly increases for unconsolidated beds (Hayer and Mehta 1982). Clearly, evaluating the fluid mechanical processes leading to the observed time and space varying τ_b and the material characteristics giving rise to τ_{ce} is a challenging task.

Bottom stress resulting from fluid processes consists, in the streamline following coordinate system, of two components

$$\tau_b = \tau_{zx} = \overline{-u'w'} + v\frac{\partial \bar{u}}{\partial z} \tag{11}$$

Here, τ_b is a force per unit area acting in the x direction in a plane (placed in the bottom) which is perpendicular to the z axis. At the bottom, the average vertical velocity $\bar{w}$ is zero. τ_b consists of the Newtonian shear based upon viscosity, v, and the Reynolds stress, $\overline{u'w'}$, which derives from the turbulent correlations. All the fluid mechanical processes described earlier integrate to give the observered time trace of bottom shear stress. Shear stress is dynamic, varies from point to point in a body of water, and is not a property of the fluid.

Erosion flux parameterizations are also governed by physical and physico-chemical properties of the bed material. In the formulae in Table 3, these properties are lumped into the critical shear stress for erosion and the various "erodibility" coefficients. These material characterizations will very much depend upon whether the particles are cohesive or noncohesive.

Table 3.   Selected Erosion Flux Parameterizations

Formula	Reference
a. $S_E = E\left(\dfrac{\tau_b}{\tau_{ce}} - 1\right)$	Partheniades (1962)
b. $S_E = \dfrac{A'D_b\gamma_s}{t(\tau_b)}\left[1 - \dfrac{1}{\sqrt{2\pi}}\int_{-a}^{a}\exp\left[-\left(\dfrac{w^2}{2}\right)\right]dw\right]$	Partheniades (1971)
c. $S_E = C_s B\exp(-Bt)$	Scarlatos (1981)
d. $S_E = \varepsilon_0\exp\alpha\left(\dfrac{\tau_b}{\tau_{ce}} - 1\right)$	Hayter and Mehta (1982)
e. $S_E = \varepsilon_f\exp\left[\alpha(\tau_b - \tau_{ce})^{\frac{1}{2}}\right]$	Parchure and Mehta (1985)
f. $E_{net} = \dfrac{a_o}{t_d^2}\left[\dfrac{\tau_b - \tau_{ce}}{\tau_{ce}}\right]$	Ziegler and Lick (1986)

where

S_E = Entrainment and net entrainment flux
τ_b = Bottom shear stress
τ_b = Critical erosion shear stress
t, t_d = Time and deposition time
$E, A', a_o, \varepsilon_o, \varepsilon_f,$ etc. = Various measures of erodibility

Noncohesive particles act independently of their neighbor particles because the surface chemistry of the particle is benign as is the particle response to changes in water chemistry. These grains are usually silica (SG = 2.65) and are in the silt size to sand size fractions, i.e., ~31 microns or greater in sphere equivalent diameter.

Cohesive sediments are an entirely different matter. Possessing quite active surface chemistry which varies as a function of the material comprising the particle, cohesive sediments are one of, if not the most challenging aspect of the bed exchange problem. Reviews are available in the documents already cited, the introductory text by Tadros (1987) and excellent articles by P. Williams and D. Williams (1989) and D. Williams and P. Williams (1989). Clay particles primarily comprise the cohesive sediments class and generally have individual particle sizes less than four microns with some as small as one half micron. The basic structure of clay is still silica, i.e., the silica tetrahedral unit and silica octahedral unit which might also enclose an aluminum, magnesium or iron atom. These units combine into sheet-like layers to form the basic minerals. In order of most inert to active, the dominant clay minerals are kaolinite, illite, and montmorillonite.

Clay particles become cohesive through two processes; chemical cementation and the interaction of Van der Waals and electric surface forces. While the electric surface forces are weaker than Van der Waals forces, they decay slower with distance away from the particle which therefore becomes an important component of flocculation in freshwater. Chemical cementation, often derived from iron oxides, gives rise to bed armoring; an increase in material shear strength at the very surface layer of bed particles. The interaction of both these processes causes flocculation to occur. Therefore aggregates of individual clay particles are formed which can have three types of floc structures as determined by the water chemistry, clay mineralogy and stress history of the floc. The first type is the salt floc which is random particle attachment caused by Van der Waals forces in the presence of salts. The second type is a nonsalt floc (fresh water) occurring in water of low cation concentration where the electric surface charges are influential. The third floc is the dispersed structure which has clay particles in a parallel orientation. As discussed by Tasi et al. (1987) fresh water flocs often trap ambient water by encirclement, as opposed to the dendritic structures found in saltwater which have dead but not trapped zones of water. The effective particle shapes of the flocs entrap water resulting in specific gravities (1.07) and much larger effective diameters thereby resulting in dramatically altered settling characteristics. As per Tsai et al. (1987), the flocs are, by and large, aggregated by shear and disaggregated by particle collisions.

The rate of erosion is also determined by the physical properties of the sediment and the physico-chemical nature of the water. For example, the type of clay comprising the bed is important as kaolinite and illite are more erodible than montmorillite clays. The erosion of a mixture of materials (i.e., different grain sizes) decreases with increasing clay content and decreases with increasing water content. This latter distinction has been specifically incorporated in the model by Ziegler and Lick (1986), as expressed by the "time after deposition". While not explicitly included as a variable, the effect of consolidated and unconsolidated behaviors is important either in the different values for the erodibility coefficients or in different functional forms; i.e., form a. (Table 3) for consolidated sediments and forms c to f. (Table 3) for cohesive unconsolidated sediments. Form "a" is also used for all noncohesive grain sizes, i.e., sand and silt, as well.

Sedimentation Fluxes and Settling Velocity

Again, the cohesive nature of clay particles affects the sedimentation flux through the settling velocity, and Table 4 contains three of the many forms of settling velocity available in the literature. Noncohesive settling is described by

Table 4.   Summary of Velocity Formulation Settling

Formula	Reference
a. $w_s = \dfrac{8d^2}{9}\dfrac{(\rho_s - \rho_w)g}{v}$	Stoke's Law (Vanoni 1975)
b. $w_s = K_s c^{\frac{4}{3}}$	Krone et al. (1977)
c. $w_s = ad^b$	Hawley (1982)

where

d = diameter
v = kinematic viscosity
ρ_s, ρ_w = density of sediment and water
K_s, a, b = coefficients

a generalized Stokes law (Table 4) (Vanoni 1975) which is valid over the entire range of particle Reynolds Numbers (R_e) encountered in practice. The drag coefficient, C_d, is nonlinear for particle Reynolds Numbers R_e greater than 1. Rouse (1937) has compiled a plot of R_e vs C_d as a function of submerged weight which is in use today (Vanoni 1975, Figure 2.1, p. 23, Figure 2.2, p. 25). Hawley (1982), and many others have noted that settling velocities for particles with diameters less than 100 μm do not follow generalized Stokes settling. This is attributed to cohesive effects and flocculation dynamics which are assumed to control floc shape, size, density, and rate of formation, all of which determine settling velocity. Smulchowski (1917) performed some of the earliest work on floc formation while Krone (1972, Table 4b) was among the first to parameterize these effects on settling as based upon lab and field observations. Recently, Tsai et al. (1987) and Lick and Lick (1988) have provided additional insight to the fresh water case, notably the role of collision frequency in floc breakup. Hawley (1982, Table 4c) formulated a cohesive sediments settling velocity as an aggregate of existing data from a wide variety of existing ocean and lake data. The coefficients a and b are size dependent.

Sedimentation flux parameterization is more ambiguous than the settling velocity and erosion flux parameterization. For noncohesive particles, if the settling flux at the bottom is greater than the turbulent flux at the bottom, then the net flux is a sedimentation flux and the calculation or estimation proceeds.

For cohesive sediments, it has been found necessary to handle settling flux in a more complex manner than simply specifying w_s and estimating $w_s \overline{c}$ and $\overline{N}_{zo}$. The cohesive functions are formulated as source/sink terms with dimen-

Table 5.   Summary of Sedimentation Flux Parameterization

Formula	References
a. $S_D = \dfrac{-PV_s c}{h}$	R.B. Krone et al. (1977)
b. $S_D = w_s c$	Y.P. Sheng (1983)
c. $S_D = \beta_L c$	Fukuda and Lick (1980)
d. $D\dfrac{\delta c}{\delta z} = -\phi w_s c$	Uchrin and Weber (1980)
e. $\phi = \dfrac{xd}{(B+d)} - \left(u_* u_{*cr}\right)$ $\left(\dfrac{\Delta C}{\Delta t}\right)_d = \dfrac{-0.434}{2\sqrt{2\pi}\sigma_2} \exp\left(\dfrac{(t_c 2/2}{t} c_0\right) \cdot$ $\cdot\left[1 - erf\left(\dfrac{1}{2\pi}\log_{10}\left[\dfrac{\tau_b^* - 1}{4\exp(-1.27\tau_{b\ min}}\right]^{2.04}\right)\right]$	Hayter and Mehta (1982)

sions of $(MT^{-1}L^{-3})$ and not flux. Table 5 summarizes the functional forms. One of the earliest empirical cohesive parameterizations is due to Krone et al. (1977, Table 5a) and sets sedimentation as a function of not only floc settling velocity but the probability of particles remaining deposited, P, which in turn, is set equal to $(1-\tau_{cd}/\tau_{cd\cdot})$. τ_{cd} is called the critical stress required for deposition. Typically, τ_{cd} is less than τ_{ce}, the critical erosion stress which essentially says that deposition and erosion of the same class size cannot occur simultaneously. An additional implication is that it takes more stress to entrain the sediment than to keep it in suspension.

An early lumped (sedimentation plus erosion) formulation by Fukuda and Lick (1980, Table 5c) stated that the source sink term was proportional to not only $\bar{c}$ but a coefficient of proportionality β_L reflecting transport by settling, Browning motion or turbulent diffusion. Several forms of β_L were formulated, but since the method is no longer in use, the reader is referred to the papers for elaboration.

Urchin and Weber (1980, Table 5d) also developed a lumped formulation where ϕ is composed of empirical coefficients α, and β, and the fluid shear velocity (u*) and the critical shear velocity required for particle motion. Since such a strong relation between shear velocity and shear stress exists, the substitution of shear into these expressions could most likely be achieved. Either way, the measurement difficulties do not improve regardless of which formulation is used.

Hayter and Mehta's (1982) formulation is based upon the observation that the rate of deposition is dependent on $\bar{c}^*$ defined as

$$\bar{c}^* = \left(\bar{c}_o\right) / \left(\bar{c}_o - \bar{c}_{eq}\right) \tag{12}$$

Here, $\bar{c}_o$ is the original amount of sediment in suspension and $\bar{c}_{eq}$ is the concentration of sediment which remains suspended. They also went on to derive an expression for $\bar{c}_{eq}$ that is a function of bed shear stress, τ_b. Table 5e presents the formulation in which t is time, t_{50} is the time at which $\bar{c}^*$ equals 50, and

$$t_c = \log_{10}\left(t / t_{50}\right)^{1/\sigma_2} \tag{13}$$

Here, σ_2 is the log-normal standard deviation of $\bar{c}^*$ vs t. Further, τ_{cd} equals the critical shear stress for deposition, and τ_b^* is the nondimensional bottom stress, i.e., $\tau_b^* = \tau_b/\tau_{cd}$.

Clearly, these are not simple formulae and although the noncohesive case is well in hand, the cohesive functions are complex perhaps representing their attempt to encompass too many water column processes, such as the lumped (erosion/sedimentation) forms of Lick and Urchin and Webber. Clearly, much of this complexity also attempts to reflect interactions with the bed at a level more sophisticated than just whether or not the particle settles to the bottom. Therefore, "bed" models have been developed to be simultaneously solved along with the water column and exchange process components.

Bed Models

This is a modeling activity at its earliest stage of development and consequently there is considerable crudeness in the strategies. Among the many attempts at formulating bed models are the initial procedure due to Krone et al. (1977) and subsequent models by Sheng (1983), Onishi and Thomas (1984), Hayter and Mehta (1982), Hayter and Pakala (1989), Ziegler and Lick (1986, 1988) and Lee and Bedford (1987). No detailed review is offered here due to the lack of formulation consensus, however, all seem to share a few similarities and pose two very difficult questions.

In general, the models are formulated with a series of N horizontal layers consisting of at minimum one unconsolidated and one consolidated layer. More than one unconsolidated layer can exist but the degree of consolidation must increase from the top most layer to the consolidated layer. As regards the second similarity, many of the models above are distinguished by how the consolidation process is modeled; yet all model this process, the simplest measures being the time after deposition and the mass of the overburden. Nevertheless, the consolidation process must be described in each layer and dynamic exchanges of mass between layers must occur as consolidation proceeds. A final similarity is that within each layer the grain size distribution must be known such that the total mass of each class size is known in each layer.

While modeling the deposition process is fairly straightforward these bed models present two fundamental questions. The first, a conceptual issue, is as follows: each model layer is comprised of a mixture of the entire grain size distribution yet the critical shear stress for erosion for certain grain size classes within the eroding layer (the topmost one) might be exceeded while not so for the remaining class sizes in that layer. The question then becomes how to model the selective entrainment of certain particles from the mixture. One possible answer is to model several class sizes separately in each layer which is computationally expensive and ignores the sheltering interactions offered by the packing of large and small particles. A second problem is how are these bed material characteristics measured? With the bed factors controlling erosion limited to the top several centimeters of bed, and with the requirements for measuring the thickness, grain size and consolidation (pore water content) in each layer, there is as yet no instrumentation for making these measurements with the required spatial resolution.

The next section addresses the issue of how to obtain the information required to use the parameterizations summarized in this section.

DATA REQUIREMENTS
Summary

With the only exception being reported by Bedford et al. (1987), there are no direct measurements of erosional flux available, either to validate the parameterizations or use directly. Field methods for estimating these fluxes have been reviewed in a forthcoming article by Bedford (1991a) and the reader is referred to this article for the various frameworks. Essentially then the adequacy of the erosion source/sink terms must be inferred from laboratory experiments and simply used in models without in situ verification. It is the author's opinion that the in situ credibility status of these parameterizations is quite low.

The information required for use of these parameterizations is quite diverse in both what it represents and how it is to be measured, therefore there is a correspondingly wide range of in situ adaptability in this information. Table 6 contains a summary of all the data that are required to make an estimate of the net bottom flux at each time step in a water body model calculation. In this table, column number 1 indicates whether the data represent water column or bed processes and column number 2 indicates where the data are obtained. In column number 2, the three traditional sources are laboratory (L) and field (F) data or predictions made from theoretical models (T). These theoretical models might in turn require additional data, which in the case of nearbottom fluid shear stress are so important, that they have also been listed. The last column is an opinion by this author as to the credibility of these data as derived from field use or as referenced to its source. The next section addresses this assessment.

The list of variables in Table 6 is daunting, not only for the shear number and range of variables but also for the lack of field measurement techniques and data for a great number of these variables. Starting with the most robustly known and proceeding to the least well-known, the rest of this section summarizes the status of these data.

Water Column Data

With a few exceptions, the water column data are a bit easier to obtain but as shown in Lee and Bedford (1987), the fluid shear stress must be very well-known as, in conjunction with the critical erosion stress, model calculations for entrainment and resuspension are most sensitive to these values. Shear stress, also not directly measured in the field, must be inferred from boundary layer theories (Grant and Madsen 1986, Bedford and Abdelrhman 1987, among many), and in general there are three classes of them, each distinguished from the other by the strength of the various forcing functions. These three classes are: (1) current-driven, (2) gravity wave-driven, and (3) combined wave-current-driven. Mathematical details are available in the review papers cited throughout this article or examples of each class may be found in Smith (1977), Vongvisessomjai (1986), and Glenn and Grant (1987), respectively. These models estimate shear stress very well when the required data are collected consistent with the assumptions inherently limiting each model. Measurement and sampling conditions have been very thoroughly reviewed in Soulsby (1980), Grant and Madsen (1986), and Bedford (1991a). Of critical importance are requirements for inertial subrange sampling (5 Hz), boundary layer averaging length (~10 min), minimal tower tilt, three-dimensional velocity data for full resolution of wave effects; and accurate bottom roughness characterization. Table 7 contains a summary of shear stress parameterizations based on the current-only boundary layer theories, although methods "e" and "f" can be adapted to the wave-current case. It should be noted that the methods requiring a "fit" with three or more points to the theory will give the highest confidence intervals.

The other significant water column data are the grain size distribution and the concentration data. Methods for measuring these data have been reviewed in Bedford (1990). Concentration per se is not actually necessary for S_E but is needed to estimate the stratification-induced suppression of bed shear stress. While high speed concentration measurements are well in hand the estimates of grain-size distribution are not.

With the promise of in situ rapid laser based (e.g., Agrawal and Pottsmith 1989) or multifrequency acoustics based (Libicki and Bedford 1989) particle sizing devices not quite fulfilled, particle sizing is typically done by collecting samples to bring back to the laboratory for analysis. The most advanced physical collection devices are those developed by Sternberg and coworkers (University of Washington, School of Oceanography), but even the most elegant configuration of this system can obtain only seven samples during a deployment. Also, these data must be collected as frequently as once per boundary layer averaging time. Unfortunately, it is this author's opinion that the more frequently we sample for grain size the more dramatic and, therefore, influential will be its time variation. In lieu of field measurements, grain-size distribution can be a water body model prediction as long as the grain-size distributions are accurately and fully known at all boundaries including the bottom.

The critical shear stress for deposition, the remaining water column variable is, at best a guess, and at worst almost completely unknown. Limited laboratory evidence exists upon which to base its estimate (Mehta et al. 1989).

Bed Data

Three classes of bed data are required. The first is configurational or geometric information for which bathymetric surveys can provide information on stable undulations. Macrobedforms must be measured on a case by case basis. Small scale ripples are predicted in the already cited boundary layer models using submodels developed from stereo field photographed geometries (e.g., Swift et al. 1982).

Measures of susceptibility to erosion are required, i.e., the critical erosion shear stress and the erodibility coefficients. For noncohesive sediments Vanoni (1975) summarizes the critical shear values originally developed from laboratory experiments of Shield's as augmented by many authors since. The cohesive critical shear stress for erosion and its accompanying erodibility coefficients are much less well known and obtained solely from laboratory devices. The excellent conference proceedings edited by Mehta and Hayter (1989) contains a number of papers on these data as does the 1985 proceedings edited by Mehta (1985). Lick and coworkers (e.g., Lick and Lick 1988, Ziegler and Lick 1988, Tsai and Lick 1987, Fukuda and Lick 1980, and Lick 1989) have examined freshwater aspects. It should be noted that the origin of the S_E parameterization, as well as the data for critical erosion shear stress for the erodibility coefficients, are for the most part derived from rotating annular flumes. The flow condition in these tanks has been assumed to be well behaved with a log-law of the wall bottom shear which would yield predictable bottom stresses. The tanks are not nearly as well behaved as first thought and the data derived from them could be

Table 6.   Input Variables and Data Requirements

Parameterization		Input data	(1) Water column (WC) or bed data (B)	(2) Origin of data	(3) Credibility status Field/origin
S_E, Erosion flux, (Table 3)	a.	critical erosion shear stress	B	L[a]	1/4[b]
	b.	erodibility coefficient	B	L	0/2
	c.	fluid shear stress (see 4. below)	WC	F	4/DNA
	d.	volume available in bed for erosion	B	F	0/DNA
	e.	grain size distribution in bed (see 3. below)	B	F	0/DNA
S_D, Sedimen- tation flux (Tables 4 and 5)	a.	grain-size distribution	WC	F	2/DNA
	b.	settling velocity	WC	T	4/4(NC)[c] 1/3(C)[d]
	c.	critical shear for deposition	WC	L	0/1
Bed response model	a.	layer thickness	B	F	0/DNA
	b.	grain size disbribution in each layer	B	F	0/DNA
	c.	pore water content in each layer	B	F	0/DNA
	d.	critical erosion	B	L	1/4
Fluid shear stress	a.	profiles of three- dimensional velocity	WC	F	3.5/DNA
	b.	gravity wave climate	WC	F	4.0/DNA
	c.	sediment grain size	WC	F	2/DNA

Table 6.   (continued)

Parameterization	Input data	(1) Water column (WC) or bed data (B)	(2) Origin of data	(3) Credibility status Field/origin
d. sediment concentration		WC	F	4/DNA
e. bedform geometry -roughness-	B		F	3/DNA
f. micro bedform geometry	B		F	4/DNA

Key
(a) R = obtained from field data
     T = obtained from theoretical models
     L = obtained form laboratory data
(b) 4 = fully developed and verified values
     3 = field data with partial theory
     2 = selected reproducable data
            results available
     1 = minimal, site specific,
            exploratory data
     0 = no instruments or data
(c) NC= Noncohesive
(d) C = Cohesive

in some considerable error (Sheng 1989). The excellent theoretical analysis by Sheng (1989) reveals strong secondary currents and possibly cellular azimuthal counter rotating cells, all of which cast doubt on the usefulness of the coefficients derived from them. Much more thorough fluid mechanical analyses of these flumes and other surrogate devices (Tsai and Lick 1987) needs to be performed before full adaptation of the results to the field situation.

Bed model parameterizations go largely untested in the field and the input data required to initialize even the model coefficients, let alone validate the model structure, cannot be obtained with the necessary precision (Libicki and Bedford 1990). The near surface character of the bottom is marked by a very dynamic consolidation zone with extremely sharp gradients in pore water and/ or particle density occurring over distances as small as 0.25 cm. Therefore, resolution in sampling pore water, grain size, and layering thickness must be of this scale. It should be understood that the processes affecting entrainment and deposition at the surface are occurring in the top centimeter of material that is

Table 7.   Methods for Calculating Bottom Stress from Constant Stress Layer Data

Formula	Reference	Formulae	Definitions
One-point empirical[a]	Sternberg (1968,1972)	$\tau_b = C_{D100F}\, U_{100}^2$	ρ = density C_{D100} = drag coefficient estimated at 100cm off bottom $\bar{U}_{100}$ = time avg. velocity 100 cm
Two-point[b]	Caldwell and Chriss (1979)	$\tau_b = \rho u_*^2 = \rho k \dfrac{(\bar{u}_{z1} - \bar{u}_{z2})}{(1nz1 - 1nz2)}$	u_* = friction velocity velocity $\bar{u}$ = measurement at points
Eddy correlation[c]	Heathershaw (1979) Gross & Nowell (1985)	$\tau_z = -\rho\,\overline{(u'w')}$	$\overline{u'w'}$ = correlation (temporal of horizontal and vertical velocity fluctuations u' and w'
Inertial dissipation method[d]	Deacon (1959)	$\tau_b = -\rho(\varepsilon k z)^{2/3}$ $\varepsilon = \left[\dfrac{\phi(k^*)}{\alpha_3}\right]^{3/2} k^{5/2}$	z = distance above bottom ε = dissipation k* = wave-number spectrum in inertial subrange $\phi(k^*) = a_3^{2/3} K^{-5/3}$

		Data/Regression Fit to	
Log profile[e]	Grant et al. (1984)	$\log z = \dfrac{k}{2.3u_*}\,\bar{u} + \log z_0$	Z_0 = roughness length
Direct Reynolds Stress[f]	Grant and Madsen (1986)	$\tau_z = \rho\overline{u'w'}$ $= u_*^2(1-Cz/l)$	$C = 5.0$ for $z < 0.11$

[a] An often used engineering approach; requires validity of log constant stress region at 100cm which is not often true; very large error in estimates possible.

[b] Holds for smooth and rough flows; heights $z1$ and $z2$ are usually separated by one decade; requires presence of constant stress region for robust estimates of bottom shear.

[c] Very susceptible to noise from wave activity and wave-current flows.

[d] Implies production and dissipation of energy in equilibrium and wave activity is outside the inertial subrange, depends on small range of wave numbers and validity of Frozen Turbulence hypothesis, insensitive to zero shifts.

[e] Slope of data-fit line to equation gives u_*, while intercepts gives z_0. Regression coefficients of 0.997 (Grant et al. 1974) or greater are required to ensure the steady nonwave affected flow required for this technique.

[f] Requires existence of log law of wall.

continually exposed to the currents. Therefore, side scan sonar images and resulting estimates of material properties are not sufficient because their sampling range is on the order of every 10 to 15 cm or greater (Libicki and Bedford 1989, 1990). Other types of measurement devices for characterizing the bottom are summarized in the article by Libicki and Bedford (1989) and, with the exception of acoustic devices that are not yet operational at the required resolutions, a vast majority of them are invasive. For the thin bed region being sampled here, the whole notion of invasive sampling of cohesive, possibly flocculated, unconsolidated sediments by invasive instruments is suspect and possibly should be abandoned. At the very least, the act of "penetration" of cores, buckets, etc. changes the pore water content of this thin layer and most probably destroys the flocs and releases the trapped interstitial waters. Nondisruptive sampling devices are critically necessary for gathering the necessary parameterization data.

ASSESSMENT

Column 3, in Table 6, contains the author's opinion as the credibility status of each of the four classes of model parameterizations listed in the table as well as a similar score for each piece of input data. Two scores are reported for each piece of input data in the form of α/β. Score β represents the degree of credibility of the data available relative to its origin. For example, the β score for the erodibility coefficients is two because the results from the laboratory are available and reproducible for selected types of conditions. The α score for the erodibility coefficients is zero because the instruments and/or the data from the field do not exist and no field test is available.

Certainly, the most credible measurements and corresponding model predictions are the fluid shear stress values. The least credible (almost nonexistent) are those for the bed models. In this author's opinion these models have far outstripped our ability to test them in the field or validate their input components in the field. The adaptation of laboratory results to the field circumstance is particularly difficult as the simplicities invoked in the laboratory apparatus often yield results that are too simple to adapt to the field situation. The opposite can also occur; as noted earlier, even the reasonably accepted rotating flume results have been called into question because a number of flow complexities are present in the tanks that are sources of bias in previously collected data.

From the management perspective, the overall measurement and parameterization status can be summarized from a "what can we do" perspective, as in Table 8. The listed parameters are those physical parameters most often referred to in management discussions or interdisciplinary research meetings. Fluxes of all types are *very* difficult to measure while bottom geometry, concentration, and velocity at selected points can be routinely obtained. Since the instruments for measuring or inferring fluxes are in some cases nonexistent, perhaps it would

Table 8.   What can be Parameterized

Class	Subclass	Status
Bottom geometry		Field measured with sonar
Point measurements	a. Concentration	Field measured with light, acoustics-based instrumentation
	b. Velocity	Field measured with acoustics-based instrumentation
Water column mass		Measurable, but not without advanced acoustic sediment profiling devices
Fluxes	a. Horizontal sediment flux	Measured only with advanced instrumentation
	b. Erosion flux	Model inferred
	c. Sedimentation flux	Measured, but not with required frequency
	d. Shear stress	Model inferred with great care
Bed material characterization		Measured, but not without destroying the sample

be advisable for biologists and chemists to adapt their information needs more to the data that can be measured routinely at this time, while still pressing the scientific need for the more informative flux data and required instrumentation.

ACKNOWLEDGMENTS

This article was prepared with support from the U.S. Army Corps of Engineers Dredging Research Program under Contract No. DACW-39-88-K-0040. Dr. Nicholas Kraus is the contract supervisor. Additional support for this work also came from Ohio Sea Grant on Research Contract Nos. R/EM-9, R/EM-12. Both sources are very much appreciated. Finally, the author has benefited considerably from the excellent graduate students working on various aspects of this research, and the author would like particularly to thank Dr. Mohamed Abdelrhman and Ms. Deborah Lee for some of the material presented here. All of this support is very much appreciated.

REFERENCES

Agrawal, Y., and H. Pottsmith. "Autonomous Long-Term In-Situ Particle Sizing using a New Laser Diffraction Instrument," *Oceans '89* (1989).

American Society of Civil Engineers "Turbulence Modeling of Surface Water Flow and Transport, Parts I-V," *J. Hydraul. Eng.* 114:970-1073 (1988).

Baker, G. and J. Gollub. *Chaotic Dynamics*, (Cambridge, UK: Cambridge Univ. Press, 1990).

Bechteler, W. *Transport of Suspended Solids in Open Channels* (Rotterdam: A. Balkema, 1986).

Bedford, K. "In-Situ Measurement of Entrainment, Resuspension and Related Processes at the Sediment Water Interface," in *Transport and Transformation of Contaminants Near the Sediment Water Interface* (New York: Springer Verlag, 1991a).

Bedford, K. "Diffusion, Dispersion and Subgrid Scale Parameterization," in *Coastal, Estuarial and Harbour Engineering Reference Book*, (London, UK: Chapman and Hall Ltd., 1991b).

Bedford, K. "In-Situ Instrumentation for the Measurement of Entrainment, Resuspension and Related Processes at Dredged Material Disposal Sites," Report U.S. Army Corps of Engineers, Coastal Engineering Research Center, Vicksburg, MS (1990).

Bedford, K., and M. Abdelrhman. "Analytical and Experimental Studies of the Benthic Boundary Layer and their Applicability to Near-Bottom Transport in Lake Erie," *J. Great Lakes Res.* 13:628-648 (1987).

Bedford, K., O. Wai, C. Libicki and R. Van Evra III. "Sediment Entrainment and Deposition Measurements in Long Island Sound," *J. Hydraul. Eng.* 113:1325-1342 (1987).

Bennett, C. and J. Myers. *Momentum, Heat and Mass Transfer* (New York: McGraw Hill Co., 1974).

Bowden K. "Physical Problems of the Benthic Boundary Layer," *Geophys. Surv.* 3:255-296 (1978).

Caldwell, D., and T. Chriss. "The Viscous Sublayer at the Sea Floor," *Science* 205:1131-1132 (1979).

Chriss, T. and D. Caldwell. "Universal Similarity and the Thickness of the Viscous Sublayer at the Ocean Floor," *J. Geophys. Res.* 89:6403-6414 (1984).

Deacon, E. "The Measurement of Turbulent Transfer in the Lower Atmosphere," *Adv. Geophys.* 6:211-228 (1959).

Fischer, H., E. List, J. Imberger and N. Brooks. *Mixing in Inland and Coastal Waters* (New York: Academic Press, 1979).

Fukuda, M., and W. Lick. "The Entrainment of Cohesive Sediments in Freshwater," *J. Geophys. Res.* 85:2813-2824 (1980).

Glenn, S., and W. Grant. "A Suspended Sediment Stratification Correction for Combined Wave-Current Flows," *J. Geophys. Res.* 92:8244-8264 (1987).

Grant, W., and O. Madsen. "The Continental-Shelf Bottom Boundary Layer," in *Annual Review of Fluid Mechanics* (Palo Alto, CA: Annual Reviews, Inc., 1986), pp.265-306.

Grant, W., A. Williams 3rd and S. Glenn. "Bottom Stress Estimates and their Prediction on the Northern California Continental Shelf During CODE-1: The Importance of Wave-Current Interaction," *J. Phys. Oceanogr.* 14:506-527 (1984).

Gregg, M. "Diapycnal Mixing in the Thermocline," *J. Geophys. Res.* 92:5249-5286 (1987).

Gross, T. and A. Nowell, "Spectral Scaling in a Tidal Boundary Layer," *J. Phys. Oceanogr.* 15:496-508 (1985).

Gust, G. "The Benthic Boundary Layer," in *Oceanography, Vol. 3* (Berlin: Springer Verlag, 1984) chapter 8.5.

Hawley, N. "Settling Velocity Distribution of Natural Aggregates," *J. Geophys. Res.* 84,9489-9498 (1982).

Hayter, E., and A. Mehta. "Modeling of Estuarial Fine Sediment Transport for Tracking Pollutant Movement," Report UFL/COE-82/009 (1982).

Hayter, E., and C. Pakala. "Transport of Organic Contamintants in Estuarial Waters," *J. Coastal Res.*, SPI 5:217-230 (1989).

Heathershaw, A., and J. Simpson. "The Sampling Variability of the Reynold's Stress and Drag Coefficient Measurements, *Estuarine and Coastal Mar. Sci.* 6:263-274 (1978).

Hinze, J. *Turbulence* (New York: American Soc. of Civil Engineers, 1975).

Holly, F. and J. Usseglio-Polatera. "Dispersion Simulation in Two-Dimensional Tidal Flow," *J. Hydraul. Eng.* 110:905-926 (1984).

Hopfinger, E. "Turbulence in Stratified Fluids: A Review," *J. Geophys. Res.* 982:5287-5304 (1987).

Krone, R. "A Cohesive Sediment Transport Model," Report 19, U.S. Army Corps of Engineers, Vicksburg, MS (1972).

Krone, R., R. Ariathuri and R. MacArthur. "Mathematical Model of Estuarial Sediment Transport," Report D-77-12, U.S. Army Corps of Engineers, Vicksburg, MS (1977).

Lee, D., and K. Bedford. "The Development of a Multi-Class Size Sediment Transport Model and its Application to Sundusky Bay, Ohio", Report Coastal Engrg. Lab, Dept. Civil Engrg., Ohio State University, Columbus, OH (1987).

Libicki, C., and K. Bedford. "Geoacoustic Properties in the Near-Surface Sediment in Response to Periodic Deposition," in *Microstructure of Fine Grained Sediments* (New York: Springer Verlag, 1990), pp.417-430.

Libicki, C., and K. Bedford. "Remote and In-Situ Methods for Sub-Bottom Characterization," *J. Coastal Res.*, SPI5:39-50 (1989).

Lick, W., and J. Lick. "Aggregation and Disaggregation of Fine-Grained Lake Sediments," *J. Great Lakes Res.* 14:514-523 (1988).

Lumley, J. "Two-Phase and Non-Newtonian Flows," *Turbulence* (Berlin: Springer Verlag, 1978), pp.289-324.

McCave, I. *The Benthic Boundary Layer* (New York: Plenum Press, 1976).

Mehta, A. *Estuarine Cohesive Sediment Dynamics* (Berlin: Springer Verlag, 1986).

Mehta, A., and E. Hayter. "High Concentration Cohesive Transport," *J. Coastal Res.* Special Issue No. 5 (1989).

Mehta, A., and E. Hayter. "Cohesive Sediment Transport, Part I," *J. Hydraul. Eng.* 115:1076-1093 (1989).

Monin, A., and A. Yaglom. *Statistical Hydromechanics* (Cambridge, MA: MIT Press, 1975).

Nihoul, J. "Bottom Turbulence" in Proceedings of the 8th International Liege Colloquium (Amsterdam: Elsevier Scientific Publishing Co., 1977).

Nowell, A. "The Benthic Boundary Layer Sediment Transport," *Rev. of Geophys. Space Phys.* 21:1181-1192 (1983).

Nowell, A., and M. Church. "Turbulent Flow in a Depth-Limited Boundary Layer," *J. Geophys. Res.* 84:4816-4824 (1979).

Onishi, Y., and F. Thomas. "Mathematical Simulation of Sediment and Radio Nucleide Transport in Coastal Waters" Technical Report, Office of Nuclear Regulatory Research, Washington DC (1984).

Parchure, T., and A. Mehta. "Erosion of Soft Cohesive Sediment Deposits," *J. Hydraul. Eng.* 111:1308-1326 (1985).

Partheniades, E. "Erosion and Deposition," in *River Mechanics, Vol. II.* (Fort Collins, CO: Water Resources Press 1971).

Partheniades, E. "A Study of Erosion and Deposition of Cohesive Soils in Saltwater," Ph.D. thesis, University of California at Berkeley, CA (1962).

Rouse, H. "Nomogram for the Settling Velocity of Spheres," Report of the Commission on Sediment, National Research Council, Washington, D.C. (1937), pp.57-64.

Scalartos, P. "On the Numerical Modeling of Cohesive Sediment Transport," *J. Hydraul. Eng..* 19:61-67 (1981).

Sheng, P. "Consideration of Flow in Rotating Annuli for Sediment Erosion and Deposition Studies," *J. Coastal Res.* SPI 5:207-216 (1989).

Sheng, Y.P. "Mathematical Modeling of Three-Dimensional Coastal Currents and Sediment Dispersion," Report CERC-83-2 U.S. Army Corps of Engineering, Vicksburg, MS (1983).

Smith, J. "Modeling of Sediment Transport on Continental Shelves," in *The Sea, Vol. 6* (New York: Wiley Interscience, 1977), pp.539-577.

Smulchowski, Z. "Versuch einer Mathematischen Theorie der Koagulations Kinetic Kolloider Losgun," *Z. Physik Chem.* 92:129 (1917).

Soulsby, R. "Selecting Record Length and Digitization Rate for Near Bed Turbulence Measurements," *J. Phys. Ocean* 10:208-219 (1980).

Sternberg, R. "Predicting Initial Motion and Bed Load Transport of Sediment Particles in Shallow Marine Environments," in *Shelf Sediment Transport: Process and Patterns*, (Stroudsburg, PA, Dowden, Hutchinson and Ross, 1972).

Sternberg, R. "Friction Factors in Tidal Channels with Differing Bed Roughness," *Mar. Geol.* 6:243-260 (1968).

Swift, W., C. Hollister and R. Chaldler. "Close-Up Stereo Photographic of Byssal Bedforms on the Nova Scotian Continental Rise," *Mar. Geol.* 66:303-322 (1982).

Tadros, Th. *Solid/Liquid Dispersions* (London, UK: Academic Press, 1987).

Tsai, C., and W. Lick. "A Portable Device for Measuring Sediment Resuspension," *J. Great Lakes Res.* 12:314-321 (1987).

Tsai, C., S. Iacobellis and W. Lick. "Flocculation of Fine Grained Lake Sediments due to a Uniform Shear Stress," *J. Great Lakes Res.* 13:135-146 (1987).

Tennekes, H., and J. Lumley. *A First Course in Turbulence* (Cambridge, MA: MIT Press, 1972).

Thompson, J., and H. Stewart. *Nonlinear Dynamics and Chaos* (Chichester, UK: John Wiley and Sons, 1986).

Urchin, C., and W. Weber. "Modeling of Transport Processes for Suspended Solids and Associated Pollutants," in *Fate and Transport Case Studies, Modelling and Toxicity* (Ann Arbor, MI: Ann Arbor Science Publishers, 1980).

Vanoni, V. *Sedimentation Engineering* (New York: American Society of Civil Engineers, 1975).

Van Rijn, L. "Mathematical Models for Sediment Concentration Profiles in Steady Flow," in *Transport of Suspended Solids in Open Channels* (Rotterdam: A. Balkema, 1986).

Vongvisessomjai, S. "Profile of Suspended Sediment due to Wave Action," *J. Waterway, Port, Coastal and Ocean Eng.* 112:35-53 (1986).

Williams, D., and P. Williams. "Rheology of Concentrated Cohesive Sediments," *J. Coastal Res.* SPI 5:165-174 (1989).

Williams, P., and D. Williams. "Rheometry of Concentrated Cohesive Suspensions," *J. Coastal Res.* SPI 5:151-164 (1989).

Woods, J. "Parameterization of Unresolved Motions," in *Modeling and Prediction in the Upper Layers of the Ocean* (Oxford, UK: Pergamon Press, 1977), pp.188-142.
Ziegler, K., and W. Lick. "A Numerical Model of Resuspension, Deposition and Transport of Fine-Grained Sediment in Shallow Water," Dept. Mechanical and Environ. Engineering, Univ. California at Santa Barbara, CA (1986).
Ziegler, K., and W. Lick. "The Transport of Fine-Grained Sediments in Shallow Waters," *Environ. Geol. Water Sci.* 11:123-132 (1988).

4 EXCHANGE OF CHEMICALS BETWEEN LAKES AND THE ATMOSPHERE

INTRODUCTION

Measurements of chemical concentrations in the atmosphere, rainfall, lake waters, biota, and sediments have revealed that there is often appreciable transfer of chemicals between the atmosphere and lakes in both directions. Indeed chemicals may be subject to cycling with the direction of net transport changing from time to time. Lakes thus respond to changes in atmospheric concentrations, and presumably atmospheric concentrations are modified by the presence of the lake. In this chapter, we review this issue by first presenting a qualitative description of the exchange processes and discussing the factors which can influence the rates of these processes. This is followed by a more detailed review and development of quantitative expressions describing these processes and the individual partitioning phenomena. Finally, these expressions are combined to describe the overall air-lake cycling phenomena.

The importance of quantifying these exchange processes is obvious. Some chemicals such as oxygen, carbon dioxide, methane, and hydrogen sulphide, arise from, or profoundly influence, important biological processes in lakes. Some organic contaminants which are of ecological or human health concern, such as PCBs, enter and leave lakes through the atmosphere. Inorganic contaminants, such as sulphur and nitrogen oxides, emitted into the atmosphere by combustion and industrial processes, profoundly affect lake processes. Metals, such as lead, may enter the lake in aerosol form, while other metals such as mercury are subject to both deposition and volatilization as they change their speciation. Atmosphere-lake linkages one of particular importance in the Great Lakes Basin, as was first comprehensively reviewed by Eisenreich (1981) and more recently reviewed again by Hites and Eisenreich (1987) and Strachan and Eisenreich (1988). Figure 1 illustrates the importance of the atmosphere as a

ISBN 0-87371-511-X

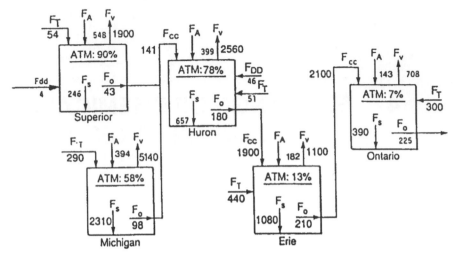

Units: kg yr⁻¹

Figure 1. **Estimated mass balance loadings of PCBs (kg/year) to the Great Lakes with the atmospheric contribution expressed as a percentage. The flows (F) are subscripted: T, tributaries; CC, connecting channels; A, atmosphere; V, volatilization; S, sediment; O, output; and DD, direct discharges. The importance of the atmosphere as a source of PCBs in the Upper Great Lakes is apparent (reproduced from Strachan and Eisenreich, 1988).**

contributor to PCB contamination of the Great Lakes. In short, the chemistry of lakes is intimately linked to the chemistry of the atmosphere above them.

The Phenomena

Figure 2 is a schematic diagram of the processes of interest, it being assumed that the exchanging chemical is present in both air and water phases.

The chemical can be present in the atmosphere in both gaseous or vapor form, and associated with aerosol particles. The concentration of these particles is normally low and varies with proximity to sources of combustion products and dust generated by wind and human activity. Concentrations range from typically 10 to 100 $\mu g/m^3$ representing a volume fraction of perhaps 5 to 50×10^{-12}. The fraction of the chemical adsorbed or absorbed (or more simply sorbed) to the aerosol particles depends on the concentration of these particles, their surface area, the properties of the chemical, and temperature. Chemicals of low vapor pressure or high boiling point tend to be most highly sorbed, presumably because they are less "soluble" in air. The most common method of measuring the

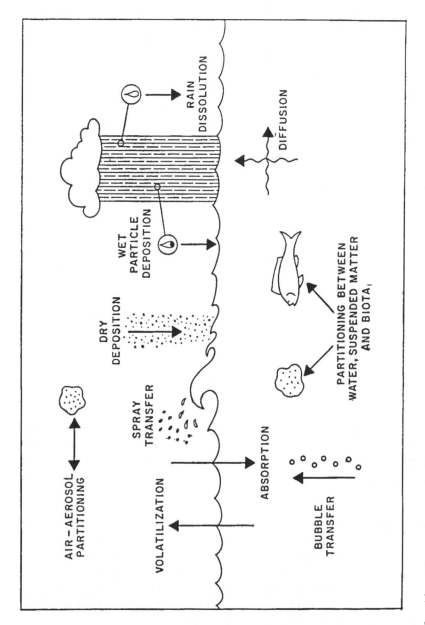

Figure 2.   Air-water exchange processes.

concentration of chemicals in vapor and sorbed form is to draw air through a filter and measure the amount of chemical trapped by the filter and present in the filtered air (Bidleman and Foreman 1987).

A similar situation exists in the water column with the chemical present in solution in water and sorbed onto suspended matter that may be mineral or organic in nature. Less soluble or more hydrophobic chemicals tend to be more highly sorbed. The concentration of suspended matter is highly variable, but is typically in the range of 5 to 50 mg/l, which for particles of unit specific gravity corresponds to a volume fraction of 5 to 50×10^{-6}, a factor of one million greater than that in the atmosphere. As with the air, the fraction sorbed and dissolved can be estimated by passing the lake water through a filter. Unfortunately, it appears that some chemical is associated with colloidal matter or very fine particles of organic matter that will pass through the filter, which usually has a pore diameter of the order of 0.2 μm. Unequivocal determination of the fractions which are sorbed and dissolved is thus very difficult, if not impossible. Elzerman and Coates (1987), Karickhoff (1984), and Hites and Eisenreich (1987) have recently reviewed these phenomena.

The interfacial region presents a particular problem because it is so difficult to sample reproducibly. There is normally a thin layer of organic material at this interface, termed the surface organic microlayer (SOML). Its organic character results in preferential partitioning of hydrophobic chemicals to this layer. Even pure air-water interfaces uncontaminated by organic matter display an excess concentration of hydrophobic chemicals, thus the interface itself appears to act as a sorbing surface. There is some controversy about the importance of this layer as influencing rates of transfer and causing biological effects, for example to surface-dwelling or surface-feeding organisms. The layer is very thin and its capacity for chemicals is low compared to the capacity of a typical meter or two depths of water column underlying it.

It is convenient to conceive of boundary layers as existing in the air and water phases within perhaps a millimeter or two of the interface. These layers consist of air and water that are relatively stagnant, that is there are few fluid eddies present that can convey chemical to or from the interface. This is in contrast to the bulk phases in the meter or so above or below the interface in which there are strong eddies or currents induced by wind or water turbulence that can efficiently convey the chemical vertically, resulting in relatively well-mixed conditions. Within the boundary layer the chemical is constrained to diffuse with a relatively low velocity controlled by the chemical's molecular diffusivity. These velocities are perhaps a few centimeters per hour in water and a few meters per hour in air but the eddies in the bulk phases well away from the interface convey the chemical at much higher velocities. The net effect is that when chemical diffuses between water and air, most of the delay, or the resistance to diffusion, occurs in these boundary layers (Mackay and Yuen 1983).

Chemical in the vapor state tends to diffuse through the air boundary layer, cross the interface, dissolve in the water, diffuse across the water boundary layer, and enter the bulk of the water. In competition with this absorption process is the reverse process of desorption, volatilization, or evaporation in which chemical in solution diffuses through the boundary layer in the water, crosses the interface, enters the air boundary phase, and diffuses through the air boundary layer to the bulk of the atmosphere. Both processes occur simultaneously. At equilibrium, the rates are equal such that there is no net transfer. The concentrations in air and water at which equilibrium is achieved are related by the air-water partition coefficient or Henry's Law constant. It is important to appreciate that only the dissolved chemical in the air and water phases is subject to this diffusion phenomenon. Sorbed chemical is not available for, or subject to, air-water diffusion.

Chemical which is sorbed to aerosol particles may fall to the water surface and become incorporated into the water column by a process called dry deposition. This process is often viewed as similar to that of a stone falling into water under the action of gravity, but this is a somewhat misleading analogy. The gravitational settling velocity of aerosol particles is very small (generally less than 1 cm/sec), thus they are conveyed freely up and down by atmospheric eddies. Statistically there is a probability that they will eventually reach the interface and become trapped. This process is nondiffusive, i.e., the rate of transfer depends only on the deposition characteristics of the particle. The chemical's presence does not influence the process, except that the deposition rate of chemical is obviously dependent on the concentration of chemical on the aerosol particle, which in turn is controlled by the gas-particle partitioning. It is difficult to measure this rate. The obvious approach is to measure the amount of chemical accumulating on a flat surface such as a glass plate coated with a sticky liquid, such as glycerol. There is some uncertainty that the presence of the plate modifies the atmospheric eddies, thus giving a measured deposition rate that differs from that which would occur over a fluid water surface. Convenient accounts of these phenomena have been compiled by Bidleman (1988) and by Pruppacher et al. (1983).

Chemical associated with aerosol particles may also be scavenged by entrapment by raindrops. This process of wet particle deposition obviously occurs only during rainfall and the rate of transport is influenced by the concentration of chemical on the aerosol, and the intensity and duration of the rainfall. The importance of this process is apparent from the increase in atmospheric visibility after rainfall scavenges particles from the air, and by the quantity of dirt present in collected rainfall. This process is also nondiffusive.

Chemical may also dissolve in rainwater during descent of the raindrop, tending to create a solution in equilibrium with the atmospheric concentration. This nondiffusive process of wet dissolution deposition is normally most important for chemicals that are highly water soluble or, more correctly, have very low Henry's Law constants or air-water partition coefficients. It is relatively easy to collect and analyze rainwater, thus determining the rate of wet deposition by both particle and dissolved routes. Usually the particle deposition route dominates.

Finally, there is a possibility that chemical present in water will be conveyed to the atmosphere in water droplets formed by breaking waves, by wind shear or by bursting bubbles. The quantity of water conveyed to the atmosphere in this way is usually small, except under very stormy conditions or near waterfalls. It is suspected that chemical associated with the SOML may be preferentially subject to this transfer process.

In summary, there are two diffusive transfer processes (absorption and volatilization) that are thermodynamically driven and that will always tend to bring the atmosphere and lake water to chemical equilibrium. There are four nondiffusive processes in which the chemical is inadvertently conveyed by being "piggy-backed" on water or aerosol which for reasons unrelated to the presence of the chemical, is journeying between air and water phases.

In addition to these transfer processes, the chemical may be introduced into the air or water region adjacent to the interface by point or nonpoint source discharges, it may be degrading by processes such as biodegradation, photolysis, or hydrolysis in both media, and it may be scavenged from the water column by sedimenting particles. The concentrations established in the few metres of air and water adjacent to the interface are thus determined by the rates of these various processes as influenced by partitioning in both phases. It is often not clear which processes are most important, and indeed in which direction net movement of the chemical is occurring.

Partitioning Equilibrium

Figure 3 illustrates the three key partitioning processes that control air-water exchange of organic contaminants. The usual approach is to estimate on a chemical-by-chemical basis the ratios of concentrations (i.e., partition coefficients) between the four media of air, water, aerosol, and suspended matter. These partition coefficients are dependent on the nature of the chemical, the nature of the sorbing phases, temperature, and the presence of dissolved matter in the water column. Partitioning principles have been reviewed by Mackay (1991).

Air-Water

Air-water equilibria are normally expressed by a Henry's Law constant or air-water partition coefficient. Henry's Law suggests that at equilibrium the partial pressure of chemical in the vapor phase is linearly related to the water concentration, the proportionality constant being a Henry's Law constant, namely

$$P = HC$$

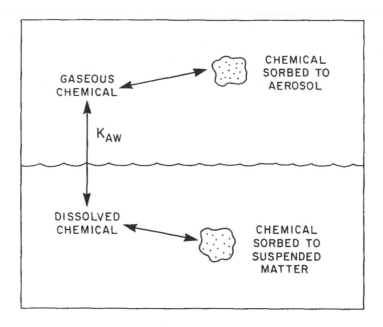

Figure 3.   Partitioning processes.

Various units may be used for partial pressure and concentration resulting in a variety of units for Henry's Law constant. For environmental purposes the preferred SI units are Pascals for pressure and mol/m³ for concentration, resulting in Henry's Law constants with units of Pa · m³/mol. The concentration in the air phase in mol/m³ can be obtained from the partial pressure by invoking the ideal gas law, namely

$$C_A = n / v = P / RT$$

It follows that a dimensionless air-water partition coefficient is given by

$$K_{AW} = C_A / C_W = P / RTC_W = H / RT$$

The magnitude of K_{AW} varies from large values of about 24 for oxygen to around unity for volatile organic chemicals, such as the halomethanes, to low values of 10^{-2} for PCBs, to 10^{-4} for polyaromatic hydrocarbons, and essentially zero for involatile metals and ionic species. In principle H or K_{AW} can be measured by bringing air and water into equilibrium with some chemical present and measuring the concentration of chemical in both phases. In practice other methods may be used such as dynamic stripping techniques. A common assumption is that Henry's Law applies not only to solutions, but also to saturation conditions. Thus H is not only P/C it is P^s/C^s where superscript s denotes saturation conditions in which equilibrium is achieved with pure chemical. P^s is then the vapor pressure of the chemical which can be obtained from tables or estimated from boiling point, and C^s is the water solubility of the chemical which can be obtained from reported data, or in some cases estimated. There are pitfalls in calculating H. It is essential that the vapor pressure and the solubility refer to the same phase, i.e., solid or liquid. This approach breaks down for liquid chemicals such as alcohols in which water has a very high solubility. It should be noted that Henry's Law constants are quite temperature sensitive, primarily because vapor pressure is a strong function of temperature. The temperature coefficient of the Henry's Law constant is the enthalpy of phase transition from solution to vapor state.

A notable feature of the Henry's Law constant is that it may retain a fairly large value for organic substances which have very low vapor pressures and therefore should not at first sight be subject to volatilization. The reason for this is that as organic molecules increase in molecular weight, for example by adding chlorines to the biphenyl molecule in the series of PCBs, the vapor pressure is reduced as expected, but the water solubility is also reduced thus the Henry's Law constant remains relatively unchanged. Highly hydrophobic substances such as PCBs, DDT, and pesticides such as chlordane or toxaphene can thus have relatively high Henry's Law constants and will be subject to appreciable volatilization from water despite the fact that they are quite involatile and have very low vapor pressure. Colloquially, the reason for this is that although the molecules certainly resist transfer to the atmosphere as manifested by their low vapor pressure, they also equally resist being dissolved in water, i.e., they are "water-hating" or hydrophobic.

A convenient review of Henry's Law constants for substances of environmental interest and the discussion of some of the theoretical basis for H are given by Mackay and Shiu (1981).

Water-Particle

Partitioning between dissolved state in water and particulate matter in water column is more complex and less well quantified. For organic chemicals, it appears that the primary sorbing sites on particles are organic matter such as humic or fulvic acids, humin, or lipid material in aquatic organisms. A common descriptor of the extent of partitioning of chemicals between organic media and water is the octanol water partition coefficient for which extensive data tabulations are available, an example being the book by Hansch and Leo (1979). K_{ow} is also important in pharmacology, and extensive efforts have been devoted to measuring it and to developing techniques for calculating it from molecular structure. K_{ow} is inversely related to solubility and indeed several correlations exist between K_{ow} and solubility. Studies of sorption between water and suspended particles demonstrate that particle water partition coefficients can be estimated by assuming the organic matter in the particles to behave somewhat similarly to octanol. The simplest approach is to assume that the organic matter is equivalent to octanol, as has been suggested by DiToro (1985). Other studies suggest that to obtain the "equivalent" amount of octanol, the mass of organic carbon should be multiplied by a factor of 0.4 to 0.6 (Karickhoff 1981). It appears that organic carbons differ in their properties and structures and thus in sorbing capacity. Therefore this approach must be used with caution (Gauthier et al. 1987).

Air-Aerosol

Characterization of the extent of sorption is normally approached by calculating the quantity $C_v \cdot TSP/C_p$ where C_v and C_p are the vapor and particulate concentrations (ng/m^3 air) and TSP is the total suspended particulate concentration (ng/m^3). This group which is a function of chemical properties (notably vapor pressure and temperature) has been used by Yamasaki et al. (1982) and Bidleman and Foreman (1987) as a basis for correlation equations. Ligocki and Pankow (1989) and Pankow (1987) have compiled excellent reviews of this issue. In the absence of experimental measurements, it is possible to discriminate between gaseous and sorbed chemical and by using correlations of these types. At lower temperatures, and for compounds of high molecular weight, the magnitude of the group decreases and the chemical becomes increasingly particle associated. Equipartitioning tends to occur for chemicals with vapor pressures of about 10^{-4} Pa.

In summary, when seeking to understand atmosphere-lake exchange processes it is first necessary to establish the sorbing properties or conditions in both phases, then develop expressions describing the rates of diffusion by volatilization or the reverse absorption and by bulk phase transfer through wet and dry deposition. Methods by which these phenomena are quantified are described in the next section.

Quantifying Air Water Exchange Processes

Methods of characterizing air-water exchange processes quantitatively are best illustrated by a calculation. The situation treated is hypothetical but resembles the exchange of PCBs between the atmosphere and a small lake of area one square kilometer. The example presented is similar to that described by Mackay et al. (1986) and employs principles reviewed by Mackay (1991). The chemical is a nonreactive organochlorine substance similar in properties to a PCB with a molecular mass of 350 g/mol. Its vapor pressure is 0.005 Pascal or 3.7×10^{-5} mm Hg and its solubility in water is 0.035 g/m^3 or 0.0001 mol/m^3. The Henry's Law Constant or a ratio of vapor pressure to solubility is thus 50 Pa·m^3/mol which corresponds to an air-water partition coefficient at 15°C or 288 K of 0.021, i.e., the water concentration at equilibrium is some 50 times that in the air.

The approach taken is to assume that certain experimentally measured concentrations of the condition of the chemical in the atmosphere and water are available which result in an ability to describe the partitioning of the chemical. From this we can deduce the rates of air-water exchange by the various processes.

The total air concentration is 6 ng/m^3 consisting of 5.66 ng/m^3 in gaseous and 0.34 ng/m^3 in particle-associated or aerosol form, i.e., 5.7% is sorbed. These data can be obtained by filtering the air. The total suspended particulate concentration is 100 µg/m^3, thus the concentration of chemical on the particles is 0.34 ng/100 µg or 3,400 ng/g. An aerosol to gas partition coefficient or ratio of concentrations can be calculated resulting in a value of about 5.7×10^8 l/kg, i.e., 3.4×10^6 ng/kg divided by 6×10^{-3} ng/l.

It is possible to calculate the air to aerosol partitioning using correlations of the Yamasaki-Bidleman type or to make direct measurements by filtering the air.

In the water column, the total concentration is 1 ng/l or 1000 ng/m^3 consisting of 803 ng/m^3 in solution and 197 ng/m^3 sorbed onto particles. The suspended particulate concentration in the water is assumed to be 12 g/m^3 thus the concentration of chemical particles is 197/12 or 16·4 ng/g. This extent of partitioning could be estimated from the organic-carbon partition coefficient K_{oc} which for a chemical of K_{ow} equal to 1,000,000 could be about 410000 thus for 5% organic carbon in the suspended matter a K_p of 0.05×410000 or 20,500 l/kg would be expected. Partitioning could thus be calculated as shown below for 1 cubic metre of water which contains a total of 1000 ng of chemical in dissolved (C_D) and sorbed (C_S) form.

$$1000ng = (1000l) \times C_D + (0.012kg) \times C_S$$

with C_D in units of ng/l, and C_S in ng/kg. But

$$C_S / C_D = 20500l / kg = K_P$$

therefore,

$$C_D = 1000 / (1000 + 0.012 \times 20500) = 1000 / 1246 = 0.803ng / ng / l$$

Again, if experimental data are not available, the relative proportions of dissolved and particulate chemical can be quantified using the approach above. However, it must be emphasized that the results could be in substantial error.

Having established the condition of the chemical in the two bulk phases it is now possible to explore their equilibrium status, i.e., how the dissolved chemical concentrations in air and water relate to the air-water partition coefficient. The water concentration of 803 ng/m^3 would be in equilibrium with an air concentration of $803 \times K_{AW}$ or 16.8 ng/m^3. This is higher by a factor of three than the "dissolved" concentration in the air, thus the water is "supersaturated" with respect to the air. Diffusive transfer, which always tends to result in an approach to equilibrium, will result in water to air diffusion, i.e., the rate of volatilization will exceed the rate of absorption.

Rates of Exchange

The rate of volatilization between water and the atmosphere (or the reverse absorption) is normally described using the two film theory of mass transfer as introduced to environmental conditions by Liss and Slater (1974) and described more recently by Mackay et al. (1986) and Mackay and Yuen (1983). The key rate determining parameters are the mass transfer coefficients for the water and air phases. These are essentially the velocities with which material can diffuse through the stagnant films on either side of the interface. They can be regarded as ratios of a diffusivity to a diffusion path length as expressed by Fick's First Law of Diffusion. These coefficients are dependent primarily on the state of turbulence existing at the water interface as influenced by wind speed and water currents. They are also temperature-dependent and are influenced by the diffusivity of the chemical. Typical values are 2×10^{-5} m/sec for k_w, the water side coefficient and 5×10^{-3} m/sec for the air side coefficient k_λ. The flux of chemical through the water layer is thus given by the expression below where C_{wi} (g/m^3) and C_w are the concentrations in the water phase at the interface and in the bulk water and A is the area (m^2)

$$N(g/sec) = k_w \cdot A(C_w - C_{wi})$$

Similarly in the air phase, the interfacial concentration is C_{ai} and the bulk phase value is C_a thus

$$N = k_\lambda A\left(C_{ai} - C_a\right)$$

Under steady-state conditions, with no accumulation of chemical at the interface, these two fluxes must be equal. Further, it can be assumed that because the air and water phases on either side of the interface are in intimate contact, equilibrium is established, thus C_{ai} and C_{wi} are related by the partition coefficient K_{AW}, namely

$$C_{ai} / C_{wi} = K_{AW}$$

Rearranging these equations to eliminate the interfacial concentrations results in the overall flux equation as shown below.

$$N = K_w A\left(C_w - C_a / K_{AW}\right)$$

where

$$1 / K_w = 1 / k_w + 1 / k_a K_{AW}$$

The term K_w is an "overall" water side mass transfer coefficient which is a function of the two individual coefficients and the air water partition coefficient. The term $(1/K_w)$ can be viewed as the total resistance to transfer which is the sum of the resistance in the water $(1/k_w)$ and the resistance in the air $(1/k_{aw}K_{AW})$. In this example, substituting the dissolved concentrations the assumed mass transfer coefficients and the area gives a flux of 8930 ng/s or 282 g/year.

$$k_w = 2 \times 10^{-5} \text{ m/sec}$$
$$k_a = 5 \times 10^{-3} \text{ m/sec}$$
$$K_{AW} = 0.0209$$

$1/k_w = 1/k_w + 1/k_A K_{AW}$ $= 50000$, i.e., 84% + 9570, i.e., 16%
  $= 59570$, i.e., 100%

$K_w = 1.68 \times 10^{-5}$ m/sec
$N = K_w A(C_w - C_a/K_{AW}) = 1.68 \times 10^{-5} \times 10^6 (803 - 5.66/0.0209)$
  $= 16.8 \times (803 - 271) = 8940$ ng/sec $= 282$ g/year

Clearly 84% of the resistance to transfer lies in the water phase and 16% in the air phase. This distribution of resistance is profoundly affected by K_{Aw}. For chemicals of very low K_{Aw}, such as polynuclear aromatics, the air phase resistance term dominates and it may then be unnecessary to calculate the water phase resistance term accurately. Similarly for volatile chemicals, with high $K_{\lambda w}$ values, detailed knowledge of k_λ may not be required.

It is important to appreciate that the net upward flux of approximately 8900 ng/sec actually represents the difference between an actual upward flux of 13500 and a downward flux of 4600 ng/sec. For a more detailed discussion of air-water diffusive exchange the reader should consult the review by Liss and Slinn (1983) that documents laboratory and environmental studies, and consult the results of the compilation by Brutsaert and Jirka (1984).

Dry Deposition

Dry deposition rates are normally characterized by a dry deposition velocity V_D which typically has a magnitude of about 0.3 cm/sec. The rate of deposition of particles is then the product of this velocity, the area, and the concentration of particles in the atmosphere, i.e.,

$$Rate = V_D . A . C_P$$

The implication is that a volume of the atmosphere $V_D A$ is depleted of particles and the associated chemical every second. In this case with V_D of 0.003 m/sec and an area of 10^6 m^2 it amounts to 3000 m^3/sec. Since each cubic meter contains 100 µg of suspended paticulates, this results in deposition of 3×10^5 µg/sec or 0.3 g/s. Since each gram of particulates is associated with 3,400 ng of chemical, this corresponds to 1020 ng/sec or 32 g/year deposition of chemical. Alternatively, and more simply, this rate can be calculated as a product of the deposition velocity, area, and the particle-associated concentration of chemical in the atmosphere, namely, 0.34 ng/m^3, i.e., $0.003 \times 10^6 \times 0.34 = 1020$ ng/sec.

Dry deposition velocities vary with aerosol particle characteristics, particularly diameter, and with the nature of the windfield as reviewed as by Slinn (1983). Very few measurements have been made of dry deposition velocities over lake surfaces. Most of the information available is for terrestial soil, or agricultural surfaces.

Wet Deposition

Wet deposition of particle-associated chemical is normally estimated by using a wash-out ratio or scavenging ratio. It is observed that each volume of rain scavenges particles from approximately 200,000 times its volume of atmosphere, this quantity being termed a *scavenging ratio*. Its magnitude is variable depending on the nature and duration of the rainfall and the efficiency with which the rain droplets scavenge the aerosol particles. Each cubic meter of water will thus, in this case, remove approximately $200,000 \times 100$ µg or 20 g of particles from the atmosphere. Interestingly, the rain contains a higher concentration of solid particulate matter than the lake water and represents a very efficient means of scavenging particles from the atmosphere, hence the observation that visibility is greatly improved after rainfall. In this case, associated with the 20 g of particles will be 68,000 ng of chemical. Thus the concentration of chemical in the rain is expected to be 68,000 ng/m^3. Assuming, for illustrative purposes, an annual rainfall of 0.5 m/year, or 500,000 m^3 over a lake area of 10^6 m^2, the annual quantity of chemical deposited by rain will be 34 g/year in association with 10 tons of particles.

Another approach is to use a chemical-specific "wash-out ratio" which is the ratio of the observed total concentration of chemical in the rainfall to total concentration in the atmosphere. In this case the wash-out ratio would be 68,000:6 or 11,300. Occasionally these wash-out ratios are also expressed on a mass concentration basis, e.g., nanograms per kilogram in each phase resulting in a value lower by the ratio of the density of water to that of air, i.e., about 800.

Also associated with the rainfall is a small quantity of chemical which may be dissolved in the water to achieve equilibrium with the air. Its concentration will be C_a/K_{AW} or 270 ng/m^3 which adds only approximately 0.5% to the concentration of chemical in rainfall which is particle associated. Only for very soluble chemicals such as alcohols is this process significant.

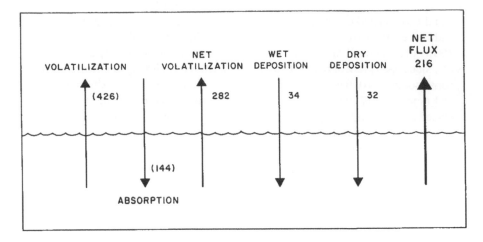

Figure 4.   Summary of estimated mass fluxes (g/year) for a hypothetical chemical through a lake-atmosphere interface.

A different and more difficult and uncertain set of calculations must be undertaken if the precipitation is snow, the scavenging or wash-out ratios being very uncertain for these media.

The overall chemical flux situation is summarized in Figure 4. The dominant process is volatilization of 282 g/year which in reality is the result of volatilization of 426 g/year and absorption of 144 g/year. Countering this process is wet deposition of 34 g/year and dry deposition of 32 g/year resulting in a net lake to atmosphere flux of 216 g/year. Clearly there is appreciable cycling of chemical between the air and water. The volatilization and dry deposition processes are relatively steady but the wet deposition process is highly intermittent, consisting of periods of intense downward movement at a rate possibly 20 times that calculated on the average.

It is apparent that a steady-state condition can be reached in which there is no net air to water flux, but the air and water are not in thermodynamic equilibrium. This occurs when the nondiffusive wet and dry deposition processes are balanced exactly by the diffusive volatilization process. It is expected that in regions remote from direct emissions of chemicals, lake waters will tend to become supersaturated with respect to the atmosphere. This condition of supersaturation will be most pronounced for chemicals which have very low vapor pressures, are aerosol associated and susceptible to wet and dry deposition, and which have high octanol-water partition coefficients thus becoming particle-associated in the water column and not available for volatilization. Low temperatures will enhance this process because they reduce the vapor pressure and Henry's Law constant, thus enhancing particle association and reducing volatilization.

It is possible, and even probable, that the net direction of air-water exchange varies seasonally with a tendency for net deposition in colder months and net volatilization in warmer months. Measurements or calculations of the rates of air-water exchange under one set of environmental conditions, such as summertime, must then be extrapolated to other conditions, such as winter, with extreme care. Indeed it may be necessary to make periodic measurements during the entire year to obtain a reliable picture of the annual flux between a specific lake and its atmosphere.

CONCLUSIONS

The processes of exchange of organic chemicals between lake water and the atmosphere have been described qualitatively and methods of quantifying these processes outlined. Obtaining a reliable picture of the air-water exchange processes requires recognition that there is often active cycling of chemicals between air and water in both directions with the net direction not necessarily being obvious and possibly changing with season. Complete quantification of the fluxes applying to a particular lake requires a detailed knowledge of the physical-chemical properties of the substance, the conditions of the atmosphere and lake water, the volatilization mass transfer coefficients, the precipitation characteristics, and how these quantities vary with season and temperature. Only then can the air-water fluxes be reliably determined and incorporated into general models describing the dynamics of chemicals in the lake as they partition from the water to biotic and abiotic phases in the water column and in the bottom sediments.

REFERENCES

Bidleman, T.F., and W.T. Foreman. "Vapor Particle Partitioning of Semivolatile Organic Compounds," in *Sources and Fates of Aquatic Pollutants. Advances in Chemistry Series 216* (Washington, D.C.: American Chemical Society, 1987).

Bidleman, T.G. "Atmospheric Processes," *Environ. Sci. Technol.* 22:361-367, and correction 22:726-727 (1988).

Brutsaert, W., and G.H. Jirka. *Gas Transfer at Water Surfaces* (Dordrecht, Holland: D. Reidel, 1984).

DiToro, D.M. "A Particle Interaction Model of Reversible Organic Chemical Sorption," *Chemosphere* 14:1503-1538 (1985).

Eisenreich, S.J. *Atmospheric Pollutants in National Waters* (Ann Arbor, MI: Ann Arbor Science, 1981).

Elzerman, A.W., and J.T. Coates. "Hydrophobic Organic Compounds on Sediments: Equilibrium and Kinetics Sorption," in *Sources and Fates of Aquatic Pollutants. Advances in Chemistry Series 216*, (Washington, D.C.: American Chemical Society, 1987), pp. 263-318.

Gauthier, T.D., W.R. Seltz and C.L. Grant. "Effects of Structural and Compositional Variations of Dissolved Humic Materials on Pyrene K_{oc} Values," *Environ. Sci. Technol.* 21:243-248 (1987).

Hansch, C., and Leo, A. *Substituent Constants for Correlation Analysis in Chemistry and Biology* (New York: Wiley - Interscience, 1979).

Hites, R.A., and S.J. Eisenreich. *Sources and Fates of Aquatic Pollutants. Advances in Chemistry Series 216* (Washington, D.C.: American Chemical Society, 1987).

Karickhoff, S.W. "Semi Empirical Estimation of Sorption of Hydrophobic Pollutants on Natural Sediments and Soils," *Chemosphere.* 10:833-849 (1981).

Karickhoff, S.W. "Organic Pollutant Sorption in Aquatic Systems," *J. Hydraul. Eng. ASCE* 110:707-735 (1984).

Ligocki, M.P., and J.E. Pankow. "Measurements of the Gas/particle Distributions of Atmospheric Organic Compounds," *Environ. Sci. Technol.* 23:75-83 (1989).

Liss, P.S., and W.G.N. Slinn. *Air-Sea Exchange of Gases and Particles* (Dordrecht: D. Rudel, 1983).

Liss, P.S. "Gas Transfer: Experiments and Geochemical Implications," in P.S. Liss and W.G.N. Slinn, Eds., *Air-Sea Exchange of Gases and Particles* (Dordrecht: D. Rudel, 1983), pp. 241-298.

Liss, P.S., and P.G. Slater. "Flux of Gases Across the Air-Sea Interface," *Nature (London)* 247:181-184 (1974).

Mackay, D., S. Paterson and W.H. Schroeder. "Model Describing the Rates of Transfer Processes of Organic Chemicals between Atmosphere and Water," *Environ. Sci. Technol.* 20:(8):810-816 (1986).

Mackay, D., and W.Y. Shiu. "A Critical Review of Henry's Law Constants for Chemicals of Environmental Interest," *J. Phys. Chem. Ref. Data.* 10:1175-1199 (1981).

Mackay, D., and A.T.K. Yeun. "Mass Transfer Correlations for Volatilization of Organic Solutes from Water," *Environ. Sci. Technol.* 17:211-216 (1983).

Mackay, D. *Multimedia Environmental Models: The Fugacity Approach* (Chelsea, MI: Lewis Publishers, 1991).

Pankow, J.F. "Review and Comparative Analysis of the Theories of Partitioning between Gas and Aerosol Particulate Phases in the Atmosphere," *Atmos. Environ.* 21:2275-2283 (1987).

Pruppacher, H.R., R.G. Semonin and W.G.N. Slinn. *Precipitation Scavenging Dry Deposition and Resuspension Vols. I and II* (New York: Elsevier, 1983).

Slinn, W.G.N. "Air-to-Sea Transfer of Particles," in P.S. Liss and W.G.N. Slinn, Eds., *Air-Sea Exchange of Gases and Particles* (Dordrecht: D. Reidel, 1983), pp. 241-298.

Strachan, W.M.J., and S.J. Eisenreich. "Mass Balancing of Toxic Chemicals in the Great Lakes: The Role of Atmospheric Deposition," Report of the International Joint Commission, Windsor, Ontario (1988).

Yamasaki, K., K. Kuwata and H. Mirgamoto. "Affects of Temperature on Aspects of Airborne Polycrylic Aromatic Hydrocarbons," *Environ. Sci. Technol.* 16:189-194 (1982).

5 MODELING THE ACCUMULATION OF POLYCYCLIC AROMATIC HYDROCARBONS BY THE AMPHIPOD *DIPOREIA* (SPP.)

INTRODUCTION

Benthic organisms are exposed to an array of chemical contaminants associated with sediments. One of the important chemical classes of sediment-associated contaminants in the Great Lakes are the polycyclic aromatic hydrocarbons (PAH) (Nriagu and Simmons 1984). Because PAH have low water solubilities and sorb to settling particles, PAH are transported rapidly to the sediments. The settling organic matter and sediment detritus in turn act as food and PAH sources for benthic invertebrates (Landrum et al. 1985, Eadie et al. 1985). High levels of PAH have been found in various benthic organisms of the Great Lakes (Eadie et al. 1985).

Among the important benthic organisms in the Great Lakes ecosystem, the amphipod *Diporeia* (spp.) represents the major benthic invertebrate on a mass basis (Marzolf 1965) and constitutes 65% of the Lake Michigan benthic fauna (Nalepa 1989). (Note: classification of the amphipod *Diporeia* (spp.) is a recent reclassification from *Pontoporeia hoyi* (Bousfield 1989). Additionally, this amphipod readily accumulates nonpolar organic contaminants from both water and sediment with no apparent biotransformation (Landrum 1988, 1989). Because of the importance of this amphipod as a major food source for many fish in the Great Lakes (Mosley and Howmiller 1977), an improved interpretation and understanding of the routes and rates of accumulation for PAH was warranted.)

Mathematical models are often developed to synthesize laboratory and field data for the purpose of understanding observed environmental processes. These models are also useful for identifying those areas that require additional study. The model described in this study examines the movement of nonpolar organic

ISBN 0-87371-511-X

contaminants, specifically PAH, into the benthic food chain, specifically the amphipod *Diporeia*. This model is a kinetics-based model that predicts the accumulation of PAH from both sediment and water. The output of the model is compared with measured concentrations of PAH in *Diporeia* and the results of sensitivity and uncertainty analyses of the model are described.

MODEL STRUCTURE

The basic structure of the model involves the calculation of the accumulation and loss of PAH congeners over time using the following rate equation:

$$dC_a / dt = K_w C_w + K_s C_s - K_d C_a \qquad (1)$$

where

K_w = Uptake clearance from overlying water (ml overlying water·g^{-1} wet weight organism·h^{-1})

K_s = Uptake clearance from sediment (g dry sediment·g^{-1} wet weight organism·h^{-1})

K_d = Elimination coefficient (h^{-1})

C_a = PAH concentration in the organism (ng contaminant·g^{-1} wet weight)

C_w = PAH concentration in the overlying water (ng contaminant·ml^{-1})

C_s = PAH concentration in the sediment (ng contaminant·g^{-1} dry weight)

Each of the rate coefficients, represented by the K values, are empirical coefficients that represent the combined effects of contaminant characteristics, organism characteristics, and environmental parameters on contaminant uptake and elimination rates. From previous studies (Landrum 1988, 1989), the most important parameters affecting the K terms include contaminant hydrophobicity, organism mass, and environmental temperature. Therefore, the model structure was designed to incorporate the time varying effects of these parameters over the course of a year.

The model selects the initial K values for a compound after the log octanol:water partition coefficient (log K_{ow}) is entered based on regression equations: $K_w =$ $-84.4(\pm30.0) + 53.5(\pm6.3)$ log K_{ow} ($r^2 = 0.95$, n = 6; Landrum 1988); log ($1/K_d$) $= 0.26(\pm0.03)$ log $K_{ow} + 0.05(\pm0.02)$ mass (mg) $+ 0.99(\pm0.22)$ ($r^2 = 0.55$, n = 73; Landrum 1988) and $K_s = 0.589(\pm0.21) - 0.479(\pm0.039)$ log K_{ow} ($r^2 = 0.9$, n = 19; Landrum 1989). Initial values for the coefficients are subsequently modified to account for changing temperature and organism mass for K_w as the simulation proceeds. The mass dependence for the coefficients was derived from measures of mass variation of the natural *Diporeia* population that was collected during the same period that the laboratory kinetics studies were performed. The mass of the animals ranged from approximately 3 mg/animal in late January increasing linearly to approximately 9 mg/animal by October and holding steady through November (Landrum 1988). This mass dependence has only been examined for the coefficients governing uptake from overlying water and elimination. The mass effect becomes statistically insignificant for K_w when the log K_{ow} is less than about four and for K_d when the log K_{ow} is less than about three (Landrum 1988). The change in K_w is evaluated as $\delta K_w = -35.3 + 9.0$ log K_{ow} for compounds of log K_{ow} greater than four. Then K_w is corrected by subtracting δK_w * mass (mg) from the initial estimate of K_w. The effect of mass on K_d is part of the initial equation.

The effect of temperature on both the uptake and elimination coefficients are then incorporated using the relationships determined from laboratory studies (Landrum 1988). The temperature profile for each day of the year is generated from an algorithm created for Lake Ontario (McCormick and Scavia 1981). The thermal correction for K_w is constant, 10.9 ml $g^{-1}h^{-1}°C^{-1}$ above or below 4°C, for all compounds with log K_{ow} greater than four (Landrum 1988). For K_d, the effects of temperature varied with season such that the K_d increased by 0.0016 $h^{-1}°C^{-1}$ above 4°C and remained constant below 4°C for Julian days 0 to 150 and 321 to 365 and increased 0.0003 $h^{-1}°C^{-1}$ above 4°C from Julian day 151 to 320 (Landrum, 1988). The thermal regime can vary with depth and provides a daily average temperature but will not provide alterations that might occur with upwelling or downwelling events.

There have not been sufficient studies to determine if K_s exhibits either mass or thermal dependence. The coefficient is assumed to exhibit both but the magnitude is unknown. For this model, we assumed that the initial sediment uptake coefficient, as determined by log K_{ow}, would vary over time in proportion to the changes that occur in K_w due to the effects of mass and temperature. The uptake from sediment was not segmented into uptake from interstitial water and particle ingestion but was instead the overall apparent first-order rate coefficient on a dry sediment basis.

The model generates a rate coefficient for each day of the year and will accommodate any number of measures of overlying water and sediment concentrations. The model uses a step approach to changes in the concentration in the water and sediment concentrations. The initially specified water or sediment concentration remains constant until a new concentration is specified for water or sediment at a specific day of the year. The new value then remains in effect until again changed. Thus, water and sediment concentrations are determined for each day of the year. The concentration in the organism is calculated by adding the net accumulation for a 24-hour period to the concentration in the organism from the previous day (Equation 1). When the model is begun, the concentration in the organism is zero. Because a zero concentration is probably not appropriate, since even neonates will have some contaminant load from their parents, the concentrations used for comparisons with field data are taken from the second simulation year. The initial values of water and sediment concentrations for julian day one were set to those measured on julian day 99 (the first day these concentrations were measured) except for benzo(a)pyrene (BaP) (Table 1). The BaP water concentration was set to that measured on julian day 109 (Table 1). The water concentration measured on julian day 99 for BaP was thought to be too large based on the concentrations measured for other PAH of similar characteristics.

MATERIALS AND METHODS
Uncertainty and Sensitivity Analyses

In an effort to quantify the model's uncertainty and to establish error bounds around the predicted animal concentration, an uncertainty analysis was used to estimate the overall variance of model output originating from the variance of uncertain model parameters. In order to determine which variables contributed most to the model's uncertainty, we first performed a sensitivity analysis on the eleven principle model parameters (Table 2). We defined the variability for the model parameters in terms of their standard errors from laboratory based regression analyses for all the terms except the initial animal mass and the effect of mass on the uptake clearance coefficient (Table 2) (Landrum 1988, 1989). For those two terms the range was limited to one standard deviation on either side of the mean value based on the mass range observed during the year (Table 2). The uncertainty associated with the measured contaminant concentration in the water and sediment due to natural variability and analytical measurement error were not included in this analysis.

To serve as a means for comparing simulation results, a base simulation was generated by setting all eleven parameters equal to their respective mean values. Then, one by one, model simulations were generated by setting each variable equal to plus or minus one standard error (or standard deviation for the mass influenced terms) of the mean values, while keeping the remaining 10 variables at their mean values (Table 2). All simulations were run for two years using phenanthrene (Phe) ($\log K_{ow} = 4.16$), and BaP ($\log K_{ow} = 6.5$). Finally, the degree of model sensitivity to each of the parameters was estimated as the absolute value of the daily percent deviation (from the base simulation) averaged over the second year of simulation for the predicted animal concentration.

Only those parameters to which the model was most sensitive (> 5% deviation from the base simulation) were used for a limited uncertainty analysis. The analysis was performed by generating a series of unique two-year model runs using all possible combinations (n = 2187) of the selected parameter settings [three settings for each parameter — minimum (mean minus SE or SD), mean and maximum (mean plus SE or SD)]. All other model components, including initial conditions, model coefficients, and hydrological conditions, were identical for each simulation. The time-varying envelope of model behavior, model uncertainty, was determined by calculating the daily mean, standard deviation, an coefficient of variation of the predicted second year animal concentrations that resulted from above simulations.

Table 1.   Concentrations of PAH Congeners in Lake Michigan Water
($ng \cdot ml^{-1}$), Sediments ($ng \cdot g^{-1}$), and *Diporeia* ($ng \cdot g^{-1}$)

Julian Day	Phe	Flanth	Pyrene	BaA	BkF	BeP	BaP
Water							
99	0.043	0.0075	0.019	0.0043	0.0237	0.0011	0.020[a]
109	0.104	0.0043	0.017	0.004	0.0032	0.0035	0.001
122	0.151	0.0203	0.018	ND	0.0046	0.0050	0.0035
170	0.177	0.0134	0.025	0.0017	0.0032	ND	0.0168
187	0.250	0.025	0.035	0.0012	0.0035	0.0059	0.0030
211	0.299	0.0262	0.036	0.0733	0.0140	0.0060	0.0216
233	0.240	0.0215	0.026	0.0029	0.0058	0.0053	0.0098
270	0.287	0.0269	0.028	ND	0.0041	0.0192	0.0069
304	0.420	0.0215	0.030	0.0018	0.0129	0.0210	0.0143
324	0.199	0.0220	0.021	ND	0.0211	0.0094	0.0034
344	0.204	0.0281	0.026	ND	0.0108	0.0106	0.0061
Sediment							
99	6	5	5	4	1	4	6
233	58	26	21	9	6	19	12
344	27	51	46	26	17	55	40
Diporeia							
99	64±23[b]	50±44	58±38	48±45	36±25	52±8	49±36
109	74±11	31±7	44±9	15±2	9±2[c]	19±22	4±1[c]
122	98±7	81±53	57±14	31±2	20±6	69±12	219±31
187	723±4[d]	167±18	192±55	95±6	105±64[d]	139[e]	92±71
211	300±95	92±17	39±6	45±7	3±1[c,d]	78±8	17±8
233	640±40[d]	56±10	35±11	31±11	172±96	57±19	94±47
270	49±18	59±9	56±13	29±4	24±1	73±9	41±11
304	43±10	49±7	57±4	30±9	11±1	119±45	73±57
324	47±16	40±6	52±6	28±5	32±10	110±25	44±6
344	52±1	65±4	74±0	53±55	18±9	73±44	28±14
CV	1–36	16–88	1–66	17–104	9–70	11–117	8–77

Note: ND = not detected, Phe = phenanthrene, Flanth = Fluoranthene, BaA = Benz(a)anthracene, BkF = Benzo(K)fluoranthene, BeP = Benzo(e)pyrene, BaP = benzo(a)pyrene, CV = range of coefficients of variation (1SD/mean) in units of percent for sediments and organisms.

Table 1. (continued)

[a] This concentration was not used in the model for simulation purposes. The concentration was judged too large for the time of year and out of line with other PAH of similar characteristics.

[b] Mean ± sd, n=3.

[c] Concentrations in question. Data are inconsistent with rest of the data for this compound.

[d] Mean ± range/2, n = 2.

[e] One sample only.

Table 2.   Observed Mean and Range of the Uncertain Model Parameters

Parameter	Minimum (–SE or SD)	Mean	Maximum (+SE or SD)	% Deviation from base BaP	Phe
Uptake rate constant from water, K_w					
K_{ow} slope term[a]	47.2	53.5	59.8	36	21
Mass term[a]	0.5	1.0	1.5	64	5
Temperature term[a]	9.4	10.9	12.4	<5	<5
K_{ow} constant term[a]	54.4	84.4	114.4	26	24
Uptake rate constant from sediment, K_s					
K_{ow} slope term[b]	0.440	0.479	0.518	<5	<5
K_{ow} constant term[b]	0.568	0.589	0.610	<5	<5
Depuration rate constant, K_d					
K_{ow} slope term[a]	0.23	0.26	0.29	41	29
Mass term[a]	0.047	0.049	0.051	<5	<5
Temperature term[a]	0.00014	0.0003	0.00046	12	3
K_{ow} constant term[a]	0.77	0.99	1.21	46	52
Initial animal mass[a]	0.2	3.2	6.2	31	25

[a] From Landrum (1988).

[b] From Landrum (1989).

Analytical Methods

Thirty liter samples of overlying water, collected 1 m above the bottom, and *Diporeia* were collected for PAH analysis approximately monthly at a site nearshore, 29 m deep in Lake Michigan. Collections were not made when prohibited by ice or weather. This nearshore station is frequently influenced by upwelling and downwelling events. Sediment samples (0 to 2 cm) were collected at the same site three times during the year: spring, summer, and fall. Filtered water samples were liquid-liquid extracted while sediments and *Diporeia* were soxhlet extracted. Yields were calculated from the addition of radiolabeled ^{3}H-BaP and ^{14}C-Phe. The PAH concentrations were measured by gas chromatography using photoionization detection for sediment, *Diporeia* and water (Table 1) (Eadie et al. 1988). The water and sediment concentrations were used to parameterize the model and the predicted concentrations in *Diporeia* were compared with the measured values.

Because the measured concentrations were at discrete times of the year and the model provides daily concentrations, the model results for a specific julian day were compared with the field results for that day. A value was calculated that would normalize each day's results, as follows:

Relative Predictability (RP) = model predicted value/field measured value

This relative predictability (RP) can then be compared between various days or averaged and compared between compounds. Values greater than one indicate an over-prediction and values less than one indicate an under-prediction.

RESULTS

The sensitivity studies indicated that the model was very sensitive to changes in the factors that dictate the uptake from water and the elimination of the compound. Changes in the terms for accumulation from the sediment environment seem to be less important. The model was particularly sensitive (percent deviation from base simulation > 5%) when subjected to maximum and minimum values of 7 of the 11 variables tested (Table 2). These included the terms for the slopes and constants of the equations used to select both the initial K_w and K_d values from K_{ow}, the mass dependence term for K_w, the temperature dependence of K_d, and the initial animal mass. The model was not significantly sensitive to the temperature dependence of K_w, the mass dependence of K_d, or to any terms in the equation describing the uptake from sediment. The greatest source of model variability for the BaP simulation was generated by the mass dependence of K_w (percent deviation from base = 63.6%). For the Phe simulation, the greatest source of model variance originated from the constant term in the depuration rate constant equation (percent deviation from base = 52.4%). The mass-dependence of the K_w term was the only parameter to show a strong compound dependence — the model was much more sensitive to this parameter for BaP (percent deviation from base = 63.6%) than it was for Phe (percent deviation from base = 5.1%).

Model prediction depends on the proper model formalism, appropriate relationships between variables, and on the appropriate parameterization. When there is uncertainty in the relationships between the variables or in the analytical results used to parameterize the model, both may cause failure of the model to predict the observed values. The uncertainty related to the analytical results was not formally introduced into the uncertainty analysis but will contribute significantly to the overall accuracy and uncertainty of the model predictions. The measurement uncertainty for the sediments and the organisms was similar (Table 1) and averaged approximately 50%. Although only single water samples were analyzed for this study, other water samples measured in our laboratory have yielded similar precision to those of the sediments and organisms. Because the water concentration is such a sensitive value, its large uncertainty has a major impact on the ability of the model to predict the observed concentrations in the organisms. In addition, frequent upwelling/downwelling events very likely affect the ambient water concentration significantly (Eadie et al. 1988).

The model uncertainty analysis produced a relatively large envelope around the predicted animal concentration for both the Phe and BaP simulations, represented as the mean of the model output plus or minus one standard deviation (Figure 1). The daily means and standard deviations were calculated from 2187 model runs (= 3^7, seven most sensitive parameters, three settings per parameter). Normalizing the daily standard deviations to the daily means produced time-varying coefficients of variation which ranged between 41% and 110% (annual mean = 62%) for the Phe simulation and between 66% and 108% (annual mean = 86%) for the BaP simulation. The coefficient of variation was negatively correlated with the mean predicted animal concentration ($r = -0.711$ for Phe, $r = -0.789$ for BaP), implying a decrease in relative model uncertainty with increasing animal body burdens.

Initial comparisons of the model results to the field measured values suggests that the model was less able to predict organism contaminant concentrations for the contaminants with smaller log K_{ow} values. For all log K_{ow} values, the RP was greater than one and in several cases the values were greater than one hundred (Figure 2). The excessive predictability, values greater than one hundred, seemed to be limited to six values: four for Phe and two for benzo(k)fluoranthene (BkF). If these values are considered outliers, most of the RP values fall below 30 (Figure 3). However, the general trend of larger RP values with smaller log K_{ow} values remains.

An improved fit of the model predictions to the field observations results when the overlying water concentration is reduced by a factor of ten. Reduction of the overlying water concentration was attempted as the first adjustment of

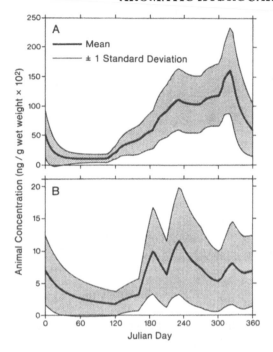

Figure 1. The envelope of the uncertainty for the predictions of Phe (a) and BaP (b) when the overlying water concentration was the measured value.

model parameters because of the large uncertainty in the analytical measurement for water and the sensitivity of the model to this parameter. This manipulation results in RP values of less than eight with most less than four (Figure 4). Again this excludes the original outliers in the data set. The model was better able to predict the *Diporeia* concentrations for the larger log K_{ow} compounds. For the case of BaP, the RP averages 1.4 ± 1.1 (mean ± sd, n = 9) and for benzo(e)pyrene (BeP) the RP averages 0.75 ± 0.45 (mean ± sd, n = 10). However, the model continued to over-predict the *Diporeia* concentrations for the compounds with smaller log K_{ow} values.

Keeping the overlying water concentration reduced by a factor of ten and examining each of the predicted and measured values permits examination of temporal patterns. Concentrations of compounds with large log K_{ow} values such as BaP and BeP are relatively well predicted, within about a factor of two, except for some values in the spring where the measured value is much larger than the predicted value (Figure 5a and b). Generally the predicted concentrations produce the same seasonal pattern as the measured concentrations. The other compound with a relatively large log K_{ow}, BkF, is not predicted well and the seasonal pattern of the predicted to the measured values seems random (Figure 5c). Concentrations of benz(a)anthracene (BaA), a compound of more interme-

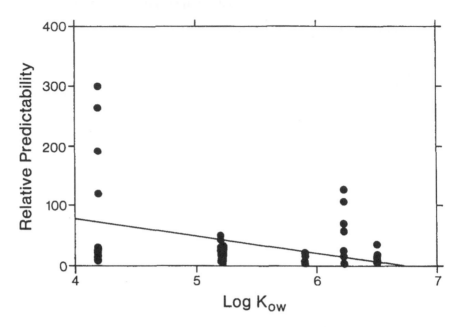

Figure 2.   Relationship between the relative predictability and log K_{ow} for all the predictions using the measured overlying water concentrations.

diate log K_{ow}, are also predicted relatively well except for one value (Figure 5d). This high predicted value for BaA is created by a single water concentration that is more than ten times larger than all the other measured concentrations. In summary, for compounds with log K_{ow} values near six, model predictions are most reasonable when the water concentration is reduced by a factor of ten. Generally, the difference between the measured and predicted values was less than a factor of two.

For compounds with log K_{ow} values about five or less, even with the water concentration reduced by a factor of ten, the predictive power of the model is limited. In general, the concentrations in the organisms are over estimated by the model simulation (Figure 5 e,f,g). For these low log K_{ow} compounds, the over-prediction is more dramatic for the later summer and fall samples compared to the winter through early summer. This generally reflects a large increase in the measured sediment concentrations in midsummer which begins to contribute substantially to the increase in the predicted organism concentrations. The measured sediment concentrations do not increase as markedly for the compounds with the larger log K_{ow} values (Table 1).

Closer examination of the flux of material into the organism for BaP, a large K_{ow} compound, shows that the predicted concentration in the organism is dominated by the flux from the overlying water year round (Figure 6a). The flux from the sediment has little impact on the predicted values even when the

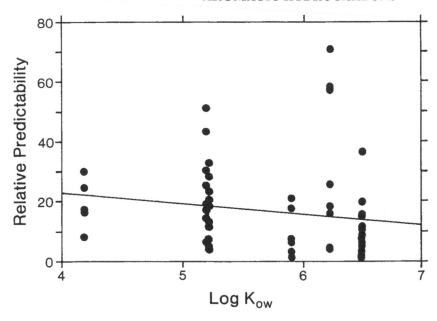

Figure 3. Relationship between the relative predictability and log K_{ow} after removing six outliers and using the measured overlying water concentrations. Relative predictability = 37.1 ($\pm$14.7) – 3.5 ($\pm$ 2.55) $\cdot$ log K_{ow} (n = 62, r^2 = 0.31, p = 0.17).

concentration in the overlying water is reduced by a factor of ten (Figure 6b). For Phe, a low K_{ow} compound, the flux from the overlying water dominates the accumulation as observed for BaP (Figure 6c) but when the water concentration is reduced by a factor of ten, the greater midsummer sediment concentrations contribute significantly to the predicted concentrations (Figure 6d). The pattern of accumulation for low K_{ow} compounds observed in the organisms is not reproduced in the model due to both higher measured sediment and overlying water concentrations in midsummer.

DISCUSSION

For a model to accurately predict the *Diporeia* concentrations, the proper model formalism, appropriate relationships between variables, and the appropriate parameterization are required. Why, then, does the model generally over-predict the concentrations observed in the field? First, the model may not have the proper formalism. The model employed for this study describes exposure to two sources, overlying water and sediment, with continuous exposure to both. These exposures may not be continuous. For overlying water, the exposures may well be discontinuous since *Diporeia* spend most of their time buried in the sediments

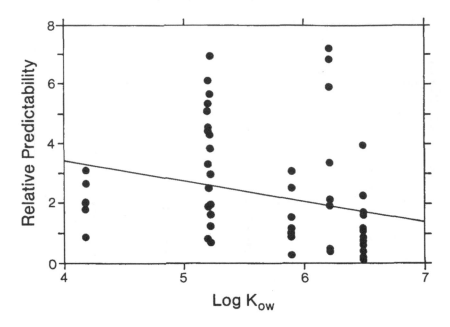

Figure 4. Relationship between the relative predictability and log K_{ow} after the overlying water concentration was reduced by a factor of ten. A relative predictability of one means the predicted value and the measured value are equal. Relative predictability = 5.5 ($\pm$ 2.1) – 0.55 ($\pm$ 0.37) · log K_{ow} (n = 62, r^2 = 0.35, p = 0.14).

with only short excursions up into the overlying water. It is estimated that these excursions account for less than ten percent of the organisms' life span (Quigly, M.A., 1990, personal communication, Great Lakes Environmental Research Laboratory, Ann Arbor, MI). Because the flux from the overlying water into the organisms results from both the concentration and the exposure time, with the current model formalism, a reduction of the concentration would be functionally equivalent to a reduced exposure duration. Reducing the simulated flux from the overlying water by decreasing the water concentration by a factor of ten seems appropriate because of the improved predictions, particularly for the compounds with larger log K_{ow} values.

The other exposure that is modeled as continuous is accumulation from sediment. This exposure does not distinguish between accumulation due to ingestion or from the interstitial water, although ingestion of sediment detrital material is thought to strongly contribute to the flux from sediment, particularly for compounds with large K_{ow} values (Landrum and Robbin 1990). Some limnological studies show that this organism seems to be an intermittent feeder (Quigley 1988) and feeds heavily on the pulse of diatom detritus in the spring,

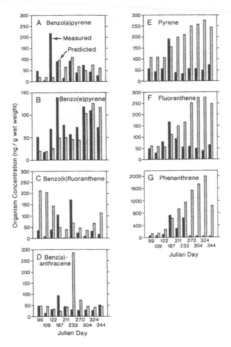

Figure 5. The comparison between the model predicted and the measured concentration of seven PAH congeners (a to g) over the course of one season.

particularly at the nearshore station that was modeled here (Gardner et al. 1985). Further, the PAH concentrations of this detrital matter are approximately a factor of five to ten greater than for the fine sediments (<53μm) (Eadie et al. 1988). Thus, freshly ingested particles should present *Diporeia* with a large pulse of contaminant in the spring and early summer when they are apparently feeding on the "raining" detrital material. Our knowledge of the model input parameters does not permit us to account for the discontinuous feeding or the potential pulse from the diatom detrital pulse. However, for compounds with large K_{ow} values, where accumulation from ingested particles is likely most important, the model yields the best predictions. Thus, not accounting for the discontinuous feeding and feeding on the spring diatoms may not be important to predicting the daily average flux of contaminant into *Diporeia*.

Finally, within the limits of the existing data, it was not possible to establish a more mechanistically based model such as an energetics model. Construction of this type of model is limited because of limited physiology, growth, and behavior data for *Diporeia* as well as constructs between these parameters and contaminant accumulation. For instance, there are only two measures of assimilation efficiency for organic compounds ingested with sediment detritus for

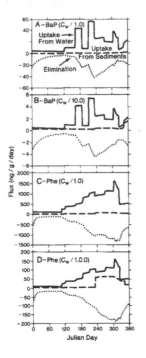

Figure 6. The fluxes for the accumulation and loss of BaP (a,b) and Phe
(c,d) over the course of one season when the overlying water
concentration was the measured values (a,c) and was reduced by
a factor of ten (b,d).

invertebrates (Lee et al. 1990, Klump et al. 1987), and these are for chlorinated
hydrocarbons and not PAH congeners. Thus, as with any modeling effort the data
limitations sometimes preclude development of more mechanistic models.
While relatively good prediction can be obtained for compounds with larger log
K_{ow} values, it is clear that improvements in predictability will require refinements
in model formalism.

The second potential reason that the model over-predicts the organism
concentrations is that the parameterization of the variables may be in error. For
instance, the actual overlying water concentrations may be lower than measured
values. There was considerable difficulty measuring the low ng l^{-1} concentrations of the individual PAH congeners and in many cases the values were near
the detection limit. The difficulty is compounded by inconsistent blanks and
potential chromatographic interferences. In a workshop conducted to balance
BaP mass in Lake Michigan, the water concentration was calculated to be
approximately 0.07 ng l^{-1} (Strahan and Eisenreich 1988). Values of PAH
measured in Lake Superior were in the range of 0.02 to 3.49 ng l^{-1} with that for
BaP at 0.4 ng l^{-1} (Baker and Eisenreich 1990). These estimates are in the range

of a factor of ten less than the values we measured (Table 1). Thus, the difficulties of making measurements of these extremely low concentrations may have resulted in over estimates of the actual water concentrations.

Further, the method of measurement, liquid-liquid extraction and gas chromatography would measure material associated with colloids or macromolecules in addition to "truly dissolved" compound. Compounds associated with these organic materials are not readily bioavailable (Landrum et al. 1987). Therefore, the measured concentrations are likely larger than the bioavailable concentrations. Finally, the temporal variability observed in the water concentrations is likely caused by frequent upwelling and downwelling that pulse local sediment resuspension. The frequency and magnitude of the true temporal variability may not be completely reflected in the data. Upwelling and downwelling events in nearshore areas occur on a more frequent basis than the sampling schedule. As a result of the wide temporal spread in field sampling, only a partial indication of the variability can be measured.

Choosing to modify the overlying water concentration, before other parameters, to obtain better model predictions is justified from the sensitivity analyses, the uncertainty about the organism's behavior, and the analytical variability. Regardless of the mechanism involved, inappropriate model formalism with regard to organism behavior or errors in analytical measures of PAH concentrations in overlying water, the organism exposure to overlying water would be one tenth that used in the initial model simulation. Reducing the exposure to PAH in the overlying water allowed the model to provide closer predictions, within a factor of about two, of PAH concentrations in *Diporeia* for compounds with a log K_{ow} approximately six. The exception in this study for compounds with relatively large log K_{ow} values was BkF and analytical interferences are suspected as the reason for the difference between model results and measurements.

Differences between the model simulation results and field results for compounds with log K_{ow} values less than six were reduced by adjusting the water concentrations. Nevertheless, there remains a trend in increasing difference between model and field measured concentrations for compounds with smaller log K_{ow} values. Thus, other explanations than reduced exposure to overlying water are required to explain why the simulations for these compounds have reduced comparability with field data. Sediment-associated PAH, particularly those with smaller log K_{ow} values, seem to become less bioavailable with increasing contact time between sediment particles and the contaminant (Landrum 1989). The process that apparently accounts for the reduced bioavailability, i.e., the movement of the compound into the pores of the particles, is likely the same process that reduces chemical extractability (Karickhoff 1980, Karickhoff and Morris 1985, DiToro et al. 1982). However, this bioavailability reduction occurs more rapidly than reduction in chemically measured concentrations (Landrum 1989). If this process occurs in the field, then the chemically measured concentrations in sediment may not reflect the biologically available contaminant. This would lead to an over estimation of the rate of accumulation from sediments for these low log K_{ow} compounds, since the mathematical formalism expects that all the compound in the sediment is bioavailable. Compounds that have larger log K_{ow} values have not shown this reduced bioavailability (Landrum 1989). Perhaps this occurs because the initial sorption is so strong that the process of migrating into particle interstices is much slower and/or the larger molecular size of the larger log K_{ow} compounds reduces their ability to migrate into pores, thus leaving the compound in a more bioavailable form. In addition to the above potential process, the rate coefficients for accumulation, K_w and K_s, behave differently for BaP and Phe as representatives of large and small log K_{ow} compounds. In the model, these rate coefficients for BaP decline in the summer by about a factor of 1.6 compared to winter values. However, the uptake rate coefficients for Phe increase during the winter to summer transition by about the same factor. These changes reflect the differences in the behavior of the uptake rate coefficients from water, K_w, for the two compounds. This difference is driven primarily by the mass dependence of K_w, which is an important variable modifying the K_w for BaP but is insignificant for K_w for Phe. Since K_s is changed in proportion to K_w, the seasonal difference in the uptake coefficients will elevate the importance of the sediment component in the accumulation of contaminant in the summer and fall, when the over-predictions seem to occur for the smaller log K_{ow} compounds. The difference is exacerbated by the increased measured values of the sediment concentrations. The combination of factors contribute to the overestimation for compounds with smaller log K_{ow} values.

The model suggests that the bulk of the *Diporeia* contaminant accumulation comes from overlying water since the larger uptake fluxes were calculated for the overlying water source. The annual estimate for the fraction of contaminant obtained from overlying water is approximately 75%, with the remaining 25% attributable to sediment uptake for both Phe and BaP when the water concentrations are held at 10% of their measured value. This proportion occurs because of the very large coefficients for the accumulation from water compared to those for accumulation from sediment (Landrum 1988, 1989). The model estimates that the relative contributions from overlying water and sediment are similar to those estimated from previous field work for Phe (Eadie et al. 1985). However, the same field study estimated that BaP would be accumulated primarily from a sediment source; the current model suggests otherwise.

The uncertainty of the model predictions is large, essentially a factor of two, from the variability of the model coefficients alone. The uncertainty contributed by the analytical variation in the parameter estimates will result in even larger uncertainty, particularly with regard to the overlying water contribution. Additional efforts will be best spent refining the analytical uncertainty in the water concentration before attempting to address the uncertainty in the model parameters.

In summary, the model markedly over-estimated the concentrations of PAH in *Diporeia* when parameterized with analytical determinations of water and sediment concentrations. The model predicted the *Diporeia* concentrations more accurately when the water concentrations were reduced by a factor of ten. Even after reducing the water concentration, the accumulation of compounds with a log K_{ow} of approximately five or less were still over estimated by up to a factor of eight. However, the compounds with a log K_{ow} of near six were generally well predicted except for BkF. After accounting for all the factors that may induce error in the predicted values, the measure of the overlying water concentrations seems the most likely parameter that accounts for most of the over-estimation.

REFERENCES

Baker, J.E., and S.J. Eisenreich. "Concentrations and Fluxes of Polycyclic Aromatic Hydrocarbons and Polychlorinated Biphenyls across the Air-Water Interface of Lake Superior," *Environ. Sci. Technol.* 24:342-352 (1990).

Bousfield, E.L. "Revised Morphological Relationships within Amphipod Genera Pontoporeia and Gammaracanthus and the "Glacial Relict" Significance of their Postglacial Distributions," *Can. J. Fish. Aquat. Sci.* 46:1714-1725 (1989).

DiToro, D.M., L.M. Horzempa, M.M. Casey and W. Richardson. "Reversible and Resistant Components of PCB Adsorption-Desorption: Adsorption Concentration Effects," *J. Great Lakes Res.* 8:336-349 (1982).

Eadie, B.J., W.R. Faust, P.F. Landrum and N.R. Morehead. "Factors Affecting Bioconcentration of PAH by the Dominant Benthic Organisms of the Great Lakes," in *Polynuclear Aromatic Hydrocarbons: Mechanisms Methods and Metabolism*, Eighth International Symposium, M. Cooke and A.J. Dennis, Eds. (Columbus, OH: Battelle Press, 1985), pp. 363-378.

Eadie, B.J., P.F. Landrum and W.R. Faust. "Existence of a Seasonal Cycle of PAH Concentration in the Amphipod, *Pontoporeia hoyi.*" *Polynuclear Aromatic Hydrocarbons: A Decade of Progress*, Tenth International Symposium, M.W. Cooke and A.J. Dennis, Eds., (Columbus, OH: Battelle Press, 1988), pp. 195-206.

Gardner, W.S., T.F. Nalepa, W.A. Frez, E.A. Cichocki and P.F. Landrum. "Seasonal Patterns in Lipid Content of Lake Michigan Macroinvertebrates," *Can. J. Fish. Aquat. Sci.* 42:1827-1832 (1985).

Karickhoff, S.W. "Sorption Kinetics of Hydrophobic Pollutants in Natural Sediments," in *Contaminants and Sediments*, Vol. 2, R.A. Baker, Ed. (Ann Arbor, MI: Ann Arbor Science, 1980), pp. 193-206.

Karickhoff, S.W., and K.P.. Morris. "Sorption Dynamics of Hydrophobic Pollutants in Sediment Suspensions," *Environ. Toxicol. Chem.* 4:469-479 (1985).

Klump, J.V., J.R. Krezoski, M.E. Smith and J.L. Kaster. "Dual Tracer Studies of the Assimilation of an Organic Contaminant from Sediments by Deposit Feeding Oligochaetes," *Can. J. Fish. Aquat. Sci.* 44:1574-1583 (1987).

Landrum, P.F., and J.A. Robbins. "Bioavailability of Sediment-Associated Contaminants to Benthic Invertebrates", in *Sediments: Chemistry and Toxicity of In-Place Pollutants*, R. Baudo, J.P. Giesy and H. Muntau Eds. (Chelsea, MI: Lewis, 1990), pp. 237-263.

Landrum, P.F. "Toxicokinetics of Organic Xenobiotics in the Amphipod *Pontoporeia Hoyi*: Role of Physiological and Environmental Variables," *Aquat. Toxicol.* 12:245-271 (1988).

Landrum, P.F. "Bioavailability and Toxicokinetics of Polycyclic Aromatic Hydrocarbons Sorbed to Sediments for the Amphipod, *Pontoporeia Hoyi*", *Environ. Sci. Tech.* 23:588-595 (1989).

Landrum, P.F., S.R. Nihart, B.J. Eadie and L.R. Herche. "Reduction in Bioavailability of Organic Contaminants to the Amphipod *Pontopreia hoyi* by Dissolved Organic Matter of Sediment Interstitial Waters," *Environ. Toxicol. Chem.* 6:11-20 (1887).

Landrum, P.F., B.J. Eadie, W.R. Faust, N.R. Morehead and M.J. McCormick. "The Role of Sediment in the Bioaccumulation of Benzo(a)pyrene by the Amphipod *Pontoporeia hoyi*," in *Polynuclear Aromatic Hydrocarbons: Mechanisms Methods and Metabolism*, Eighth International Symposium, M. Cooke and A.J. Dennis, Eds. (Columbus, OH: Battelle Press, 1985), pp. 799-812.

Lee, H., B.L. Boese, R. Randall and J. Pelletier. "Method to Determine the Gut Uptake Efficiencies for Hydrophobic Pollutants in a Deposit-Feeding Clam," *Environ. Toxicol. Chem.* 9:215-219 (1990).

Marzolf, G.R. "Substrate Relations of the Burrowing Amphipod, *Pontoporeia affinis* in Lake Michigan," *Ecology* 46:579-592 (1965).

McCormick, M.J., and D. Scavia. "Calculation of Vertical Profiles of Lake-Averaged Temperature and Diffusivity in Lakes Ontario and Washington," *Water Resour. Res.* 17:305-310 (1981).

Mosley, S.C., and R.P. Howmiller. "Zoobenthos of Lake Michigan. Environmental Status of the Lake Michigan Region," Argonne National Laboratory, ANL/IES-40, Volume 6, (1988), p.148.

Nalepa, T.F. "Estimates of Macroinvertebrate Biomass in Lake Michigan," *J. Great Lakes Res.* 15:437-443 (1989).

Nriagu, J.O., and M.S. Simmons. *Toxic Contaminants in the Great Lakes* (New York: John Wiley & Sons, 1984), p. 527,

Quigley, M.A. "Gut Fullness of the Deposit-Feeding Amphipod, *Pontoporeia hoyi*, in Southeastern Lake Michigan," *J. Great Lakes Res.* 14:178-187 (1988).

Strachan, W.M.J., and S.J. Eisenreich. "Mass Balancing of Toxic Chemicals in the Great Lakes: The Role of Atmospheric Deposition. Appendix I, The Workshop on the Estimation of Atmospheric Loadings of Toxic Chemicals to the Great Lakes Basin," International Joint Commission, Windsor, Ontario, Canada, (1988), p. 113.

6   MODELING THE ACCUMULATION AND TOXICITY OF ORGANIC CHEMICALS IN AQUATIC FOOD CHAINS

INTRODUCTION

For the management of contaminants in rivers and lakes, it is important to be able to predict the impacts of a discharge of a single and/or multiple chemicals in terms of chemical concentrations and toxic effects in aquatic biota. Building such a predictive capability requires knowledge of the transport of the chemical from its origin into the organism and of the interaction of the chemical with crucial target sites in the organism. This knowledge can then be captured in computer models, which when peer reviewed, represents the best available knowledge for practical use in water quality management. Further research will improve these models and provide a better predictive ability and management.

Several chapters in this volume discuss the factors that control the water and sediment concentrations in aquatic systems and formulate models that can be used for predicting water and sediment concentrations from chemical discharges. This chapter addresses the uptake and bioaccumulation of organic chemicals from the water and sediments into single organisms and in entire food chains. It summarizes the current state of knowledge regarding the mechanism of chemical uptake and bioaccumulation in various aquatic organisms and it presents a model to predict the accumulation of organic substances in aquatic food chains. It is further shown how this model can be used to assess toxic effects in fish and other aquatic organisms.

We will demonstrate the ability of this "food chain" model to predict chemical concentrations in aquatic food chains by applying the model to experimental data for Lake Ontario. Since the model only requires a small set of basic, and readily accessible data to characterize the food chain, the model is believed to be a practical tool for contaminant management on an "ecosystem" level.

ISBN 0-87371-511-X

BIOAVAILABILITY

One of the most important factors controlling the uptake and bioaccumulation of organic chemicals in aquatic organisms, is the concentration of the chemical in the water that can be absorbed by the organism from the water (e.g., via the gills in fish). In particular for very hydrophobic chemicals, the concentration of absorbable or bioavailable chemical is often only a fraction of the total chemical concentration in the water. This fraction is usually referred to as the Bioavailable Solute Fraction (BSF), or simply the bioavailability (Landrum et al. 1985, Black and McCarthy 1988, Gobas et al. 1989a).

The bioavailability of organic chemicals in natural waters is largely determined by the interaction of the chemical with organic carbon-containing materials, which occur both in particulate and in dissolved form. There remains controversy about the state of the chemical when sorbed to the organic matter. The most generally held current belief is that the chemical is dissolved in a solid solution form in, or on, a sponge-like matrix of organic matter, which is too large to permeate through biological membranes (e.g., the gill membrane). Sorption is generally viewed as an equilibrium partitioning of the chemical between the water and the organic matter (OM) in the water column (Karickhoff 1984)

$$\text{Dissolved Chemical} + \text{OM} \Longleftrightarrow \text{Chemical} - \text{OM} \qquad (1)$$

which can be expressed by a partition coefficient K_{oc} (l/kg):

$$K_{oc} = C_{WB} / \left(C_{WD} \cdot [OM] \right) \qquad (2)$$

where C_{WB} (μg/l) is the concentration of "bound" or "sorbed" chemical, i.e., ($C_{WT} - C_{WD}$), C_{WD} is the concentration of dissolved chemical (μg/l) and [OM] is the concentration of organic matter in the water (kg/l). Based on the hypothesis that only chemical in true solution is bioavailable, the bioavailability can be defined as the ratio of the truly dissolved chemical concentration in the water C_{WD} (μg/l) and the total chemical concentration in the water C_{WT}(μg/l), i.e.,

$$BSF = C_{WD} / C_{WT} \qquad (3)$$

Since, it has been suggested that within experimental error, K_{oc} equals K_{ow}/d_{oc} (DiToro 1985), the following expression can be derived by substituting Equation 2 in Equation 3 to estimate the bioavailability of organic chemicals in natural waters.

$$BSF = 1 / \left(1 + K_{ow} \cdot [OM] / d_{oc} \right) \qquad (4)$$

where

d_{oc} = the density of organic carbon (kg/L)

Typical concentrations of organic matter in natural waters vary between approximately 10^{-6} to 10^{-7} kg/l. If for example, the organic matter content of the water is 10^{-6} kg/l, Equation 4 predicts a bioavailability for a chemical with a log K_{ow} of 6 of 50%. This means that only half of the chemical concentration in the water can be absorbed by organisms, whereas the other half is in a nonabsorbable form.

CHEMICAL UPTAKE AND BIOACCUMULATION IN BENTHIC INVERTEBRATES

It is often convenient to view the uptake and bioaccumulation of hydrophobic organic substances in benthic invertebrates as the result of an equilibrium partitioning of the chemical between the lipids of the organism, the organic carbon fraction (OC) of the sediment, and the interstitial (or pore) water (Shea 1988, Gobas et al. 1989b).

$$C_B \cdot d_L / L_B = C_s \cdot d_{oc} / OC = K_{LW} \cdot C_P \qquad (5)$$

where C_B is the chemical concentrations in the benthic invertebrates (μg/kg wet weight), C_S is the concentration in the sediments (μg/kg dry weight), C_P is the truly dissolved chemical concentration in the pore water (μg/l water); L_B is the lipid fraction of the benthos (kg lipid/kg organism), d_L is the density of the lipids of the benthos (kg/l), OC is the organic carbon fraction of the sediments (kg organic carbon/kg organism), d_{oc} is the density of the organic carbon fraction of the sediments (kg/l), and K_{LW} is the lipid-water partition coefficient.

An interesting aspect of this model is that the organism/sediment concentration ratio C_B/C_S is only dependent on organism and sediment characteristics, namely L_B, d_L, OC and d_{oc}. The nature and properties of the chemical (e.g., K_{ow}) are not important. In other words, if the equilibrium assumption applies in the field, C_B/C_S should be approximately similar for organic chemicals, namely $L_B \cdot D_{oc}/OC \cdot d_L$, or simply L_B/OC since d_L and d_{oc} are approximately the same. If, for example, L_B is 6% and OC is 2%, then the concentration in the benthic invertebrates is approximately three times higher than the concentration in the sediments.

This simple equilibrium model ignores the physiological and time-dependent processes of chemical exchange between the organism, the sediments and the interstitial and overlying water. Alternative kinetic models, notably by Landrum (e.g., Landrum et al. 1992, this volume), achieve a greater detail by treating independently chemical uptake from water (i.e., via the gills), uptake from ingestion of sediment associated mater (i.e., via the gastrointestinal tract), and elimination to the water (i.e., via the gills), elimination into egested "faecal" matter (via the gastrointestinal tract) and elimination by metabolic transformation (i.e., for metabolizable chemicals). From a physiological point of view, this kinetic model is more correct, but it relies on laboratory measurements of several rate constants. Presently, these measurements are difficult to make, which may explain the poor agreement of the kinetic model with observed field data (Landrum et al. 1992, Gobas et al. 1989b). With continued research, the predictability of the kinetic model is likely to improve. In anticipation of better models, we adopt the simpler sediment-organism equilibrium model for benthic invertebrates. The applicability of this model can be demonstrated by comparing model predictions to field data. The results of such a comparison are listed in Table 1, which compiles results for chemicals varying in log K_{ow} from 2.7 to 8.3, for various species of benthic invertebrates, i.e., *Tubifex tubifex*, *Limnodrilus hofmeisteri*, *Pontoporeia affinis* (data from Oliver and Niimi 1988 and Fox et al. 1983), *and Hexagenia limbata* (data from Gobas et al. 1989b). The data in Table 1 indicate that chemical concentrations in benthic invertebrates and in sediments are approximately equal if they are expressed on respectively a lipid (i.e., as C_B/L_B) and on an organic carbon (i.e., as C_S/OC) basis; i.e., the benthos/sediment concentration ratio or BSR (i.e., $C_B \cdot OC/L_B \cdot C_S$) is 0.98, with a standard deviation of a factor of 3 (n = 203). This is in satisfactory agreement with a model, which predicts a value of approximately 1.0.

CHEMICAL UPTAKE AND BIOACCUMULATION IN AQUATIC MACROPHYTES

Various studies have shown that the uptake and bioaccumulation of chemicals in submerged aquatic macrophytes and phytoplankton are largely controlled by chemical exchange between the organism and the water (Geyer et al. 1984, Mallhot 1987, Gobas et al. 1991). Growth of the individual organism or population growth can also play a role, since an increase in the organism's weight or volume has a "diluting" effect on the chemical concentration in the organism. The following model represents this:

$$dC_A / dt = k_1 \cdot C_{WD} - k_2 \cdot C_A - k_G \cdot C_A \qquad (6)$$

where C_A (µg/kg) is the chemical concentration in the organism and C_{WD} (µg/l) is the bioavailable concentration in the water. k_1 (l/kg/d) and k_2 (1/d) are the first-order rate constants for respectively chemical uptake from the water and chemical elimination to the water. k_G (1/d) is the rate constant for growth, thus assuming that growth can be described by a first-order rate constant. This model has the following steady-state solution:

Table 1.   Observed Benthos/Sediment Concentration Ratios (BSR).

Tubifex tubifex and *Limnodrilus hoffmeisteri*, Lake Ontario[a]

Chemical	log K_{ow}	BSR	Chemical	log K_{ow}	BSR
PCB-28	5.80	1.11	PCB-146	6.90	0.48
PCB-18	5.60	3.78	PCB-141	6.90	0.44
PCB-22	5.60	1.89	PCB-128	7.00	0.99
PCB-26	5.50	16.20	PCB-151	6.90	0.66
PCB-33	5.80	9.72	PCB-132	7.30	0.44
PCB-17	5.60	5.40	PCB-156	6.90	0.39
PCB-25	5.50	5.40	PCB-136	6.70	1.16
PCB-24	5.50	2.70	PCB-180	7.00	0.83
PCB-32	5.80	4.73	PCB-187	7.00	1.38
PCB-66	5.80	0.49	PCB-170	6.90	0.65
PCB-70	5.90	1.10	PCB-183	7.00	1.22
PCB-56	6.00	0.53	PCB-177	7.00	0.54
PCB-52	6.10	0.68	PCB-174	7.00	0.26
PCB-47	5.90	0.81	PCB-178	7.00	0.32
PCB-44	6.00	0.81	PCB-171	6.70	0.57
PCB-74	6.10	2.20	PCB-203	7.10	0.43
PCB-49	6.10	0.86	PCB-201	7.50	0.34
PCB-64	6.10	0.69	PCB-194	7.10	0.66
PCB-42	5.60	0.69	PCB-209	8.26	0.34
PCB-53	6.10	9.45	ppDDE	5.70	0.36
PCB-40	5.60	0.96	ppDDT	5.80	0.18
PCB-46	6.00	3.60	mirex	6.89	0.60
PCB-45	6.00	0.74	γ-chlordane	2.78	0.26
PCB-101	6.40	0.77	α-BHC	3.81	1.98
PCB-84	6.10	0.63	lindane	3.80	0.27
PCB-118	6.40	0.77	HCBD	4.80	0.09
PCB-110	6.40	0.55	OCS	6.20	0.61
PCB-87	6.50	2.43	HCB	5.47	0.09
PCB-105	6.40	0.70	QCB	5.03	0.07
PCB-95	6.40	0.95	TeCB-1,2,3,5	4.65	0.04
PCB-85	6.20	0.44	TeCB-1,2,4,5	4.51	0.10
PCB-92	6.50	1.22	TeCB-1,2,3,4	4.75	0.03
PCB-82	6.20	0.47	TCB-1,3,5	4.02	0.21
PCB-91	6.30	0.57	TCB-1,2,4	3.98	0.02
PCB-99	6.60	0.98	TCT-2,4,5	4.72	0.09
PCB-153	6.90	0.81	TCT-2,3,6	4.80	0.17
PCB-138	7.00	0.56	PCT	6.20	0.27
PCB-149	6.80	0.73			

Table 1.   (continued)

Pontoporeia affinis, **Lake Ontario**[a]

Chemical	log K_{ow}	BSR	Chemical	log K_{ow}	BSR
PCB-28	5.80	1.59	PCB-170	6.90	2.34
PCB-18	5.60	2.16	PCB-183	7.00	3.77
PCB-22	5.60	1.53	PCB-177	7.00	2.12
PCB-26	5.50	22.50	PCB-174	7.00	1.66
PCB-33	5.80	5.58	PCB-178	7.00	1.80
PCB-17	5.60	4.50	PCB-171	6.70	1.94
PCB-25	5.50	5.40	PCB-185	7.00	1.53
PCB-24	5.50	5.85	PCB-173	7.00	1.97
PCB-32	5.80	3.83	PCB-203	7.10	0.55
PCB-66	5.80	0.59	PCB-201	7.50	0.51
PCB-70	5.90	1.25	PCB-194	7.10	0.88
PCB-56	6.00	0.68	PCB-195	7.10	0.98
PCB-52	6.10	0.79	PCB-206	7.20	0.09
PCB-47	5.90	1.05	PCB-209	8.26	0.07
PCB-44	6.00	1.02	PCB-31	5.70	1.59
PCB-74	6.10	2.73	PCB-27	5.80	5.85
PCB-49	6.10	1.31	PCB-76	6.00	1.25
PCB-64	6.10	0.80	PCB-60	5.90	0.68
PCB-42	5.60	0.88	PCB-81	6.10	0.68
PCB-53	6.10	10.80	PCB-48	6.10	1.05
PCB-40	5.60	0.93	PCB-97	6.60	2.75
PCB-46	6.00	5.80	PCB-182	7.00	1.50
PCB-45	6.00	1.39	PCB-190	7.00	2.34
PCB-101	6.40	1.23	PCB-196	7.50	0.55
PCB-84	6.10	1.03	ppDDE	5.70	0.64
PCB-118	6.40	1.26	ppDDD		0.07
PCB-110	6.40	0.68	ppDDT	5.80	0.43
PCB-87	6.50	2.75	mirex	6.89	0.35
PCB-105	6.40	1.08	photomirex		0.97
PCB-95	6.40	1.41	γ-chlordane	2.78	2.49
PCB-85	6.20	0.74	α-BHC	3.81	12.60
PCB-92	6.50	1.19	lindane	3.80	2.97
PCB-82	6.20	0.84	HCBD	4.80	0.12
PCB-91	6.30	0.92	OCS	6.20	0.56
PCB-99	6.60	1.08	HCB	5.47	0.16

Table 1.   (continued)

Pontoporeia affinis, **Lake Ontario**[a]

Chemical	log K_{ow}	BSR	Chemical	log K_{ow}	BSR
PCB-153	6.90	1.62	QCB	5.03	0.14
PCB-138	7.00	1.56	TeCB-1,2,3,5	4.65	0.07
PCB-149	6.80	1.22	TeCB-1,2,4,5	4.51	0.12
PCB-146	6.90	0.83	TeCB-1,2,3,4	4.75	0.17
PCB-141	6.90	1.46	TCB-1,3,5	4.02	0.11
PCB-128	7.00	0.72	TCB-1,2,4	3.98	0.09
PCB-151	6.90	1.39	TCB-1,2,3	4.04	0.18
PCB-132	7.30	0.70	TCT-2,4,5	4.72	0.50
PCB-156	6.90	1.67	TCT-2,3,6	4.80	0.70
PCB-136	6.70	7.46	PCT	6.20	0.41
PCB-180	7.00	3.32			
PCB-187	7.00	1.50			

Hexagenia limbata, **Lake St. Clair**[b]

Chemical	log K_{ow}	BSR
QCB	5.03	0.28
HCB	5.45	0.28
OCS	6.29	0.74
PCB-101	6.40	0.92
PCB-87	6.50	1.08
PCB-118	6.40	0.82
PCB-153	6.90	1.42
PCB-138	7.00	1.08
PCB-180	7.00	1.24

Tubifex tubifex and *Pontoporeia affinis*, **Lake Ontario**[c]

Chemical	log K_{ow}	BSR	BSR
TCB-1,3,5	4.02	3.34	2.11
TCB-1,2,4	3.98	2.35	1.76
TCB-1,2,3	4.04	2.09	1.86
HCBD	4.80	1.42	1.45

Table 1. (continued)

Tubifex tubifex and Pontoporeia affinis Lake Ontario[c]

Chemical	log K_{ow}	BSR	BSR
TeCB-1,2,4,5	4.51	3.31	1.79
TeCB-1,2,3,4	4.75	2.33	2.35
QCB	5.03	5.03	5.99
HCB	5.47	10.50	8.31

Note: The concentration in the benthos is expressed on a lipid weight basis, the concentration in the sediment is expressed on an organic carbon basis.

[a] Oliver and Niimi 1988.
[b] Gobas et al. 1989b.
[c] Fox et al. 1983.

$$BCF = C_A / C_{WD} = k_1 / \left(k_2 + k_G\right) \qquad (7)$$

where the ratio of the concentration in the organism and that in the water is often referred to as the bioconcentration factor BCF. If k_G is of the same magnitude or larger than k_2, the size of the growth factor k_G can have a significant effect on the bioconcentration factor. However, unless macrophyte populations are in a rapid growth phase (e.g., at certain times in the spring or summer), k_G may be small compared to k_2, which simplifies BCF to k_1/k_2. Since lipids are usually the predominant site for bioaccumulation of hydrophobic substances, BCF can be satisfactorily approximated by the chemical's octanol-water partition coefficient K_{ow}, giving

$$BCF = C_A / C_{WD} = k_1 / k_2 = L_A \cdot K_{ow} \qquad (8)$$

where L_A is the lipid content of the macrophytes (Gobas et al. 1991 for aquatic plants, Geyer et al. 1984 for phytoplankton).

CHEMICAL UPTAKE AND BIOACCUMULATION IN ZOOPLANKTON

Due to their small size and large area/volume ratio, uptake and bioaccumulation of organic chemicals in zooplankton are predominantly due to chemical exchange between the organism and the water (Clayton et al. 1977). A similar process has been discussed for aquatic macrophytes, and the following model can be proposed to estimate chemical concentrations in zooplankton C_Z (μg/kg):

$$BCF = C_Z / C_{WD} = k_1 / k_2 = L_Z \cdot K_{ow} \qquad (9)$$

where

L_Z = lipid content of the zooplankton

CHEMICAL UPTAKE AND BIOACCUMULATION IN FISH

Fish absorb chemicals directly from the water, i.e., via the gills and through the consumption of food, i.e., via the gastrointestinal tract (Bruggeman et al. 1981). Other uptake routes such as chemical absorption via the skin are usually considered to be insignificant. Chemical loss or elimination can occur via the gills to the water (i.e., essentially the reverse process of chemical uptake from the water), via egestion of faecal matter or as a result of metabolic transformation. The following equation combines these processes in an overall flux equation, describing the net flux of chemical into the fish as the sum of all of the uptake and loss fluxes:

$$d\left(V_F \cdot C_F\right) / dt = k_1 \cdot V_F \cdot C_{WD} - k_2 \cdot V_F \cdot C_F + k_D \cdot V_F \cdot C_D$$
$$- k_E \cdot V_F \cdot C_F - k_M \cdot V_F \cdot C_F \qquad (10)$$

where

C_{WD} = dissolved chemical concentration in the water (μg/l)
C_D = chemical concentration in the food (μg/kg)
C_F = chemical concentration in the fish (μg/kg fish)
V_F = weight of the fish (kg)

The 'k's are first order rate constants: k_1 for uptake from the water via the gills (l/kg·day); k_2 for elimination via the gills to the water (1/day); k_D for chemical uptake from food (kg food/kg fish/day); k_E for elimination by faecal egestion (1/day) and k_M for metabolic transformation of the chemical (in 1/day).

If fish growth is insignificant (V_F is constant), the steady-state solution of this model is

$$C_F = \left(k_1 \cdot C_{WD} + k_D \cdot C_D\right) / \left(k_2 + k_E + k_M\right) \qquad (11)$$

where

$$k_1/(k_2 + k_E + k_M) = \text{bioconcentration factor}$$
$$k_D/(k_2 + k_E + k_M) = \text{biomagnification factor}$$

Various authors have suggested that fish growth can have a considerable effect on bioaccumulation factors and concentrations in fish, in particular for chemicals of high K_{ow} (Thomann and Connolly 1984, Clark et al. 1990). From a modeling perspective, the effect of growth can be treated by varying V_F in Equation 10 with time (Gobas et al. 1989a). However, in that case a simple steady-state solution does not exist. To incorporate fish growth into the model and to allow for a simple steady-state solution to be applied, the effect of growth can be introduced in terms of a rate constant k_G, i.e., $dV_F/V_F \cdot dt$, which has units of 1/day.

If this simplification for fish growth is introduced into the fish bioaccumulation model, the flux equation becomes

$$dC_F / dt = k_1 \cdot C_{WD} - k_2 \cdot C_F + k_D \cdot C_D - k_E \cdot C_F - k_M \cdot C_F - k_G \cdot C_F \quad (12)$$

and the steady-state mass balance equation is

$$C_F = \left(k_1 \cdot C_{WD} + k_D \cdot C_D\right) / \left(k_2 + k_E + k_M + k_G\right) \quad (13)$$

where

$$k_1/(k_2 + k_E + k_M + k_G) = \text{bioconcentration factor}$$
$$k_D/(k_2 + k_E + k_M + k_G) = \text{biomagnification factor}$$

From Equation 13 it follows that to estimate steady-state chemical concentrations in fish, information is required of the values of k_1, k_2, k_D, k_E, k_M, and k_G for different chemicals and fish species.

Gill Uptake

The rate at which chemicals are absorbed by fish via the gills is expressed by the gill uptake rate constant k_1 which has units of l/kg·day. The gill uptake rate is the combined process of the gill ventilation rate G_V (m³/day) and the diffusion rate of the chemical across the gills (Gobas and Mackay 1987). The extent to which chemicals that enter the gill compartment by gill ventilation are actually absorbed by the organism is usually expressed by the gill uptake efficiency E_W. The uptake rate constant k_1 then follows as

$$k_1 = E_W \cdot G_V / V_F \quad (14)$$

Studies of the relationship between k_1 and K_{ow} and between E_w and K_{ow} in various fish species have shown that (1) k_1 and E_w increase with K_{ow} if log K_{ow} is low (< 4.5 to 5), (2) k_1 and E_w are constant if K_{ow} is large (between 5 to 7), and (3) k_1 and E_w drop with increasing K_{ow} for chemicals with extremely high K_{ow} (log K_{ow} above 7) (McKim et al. 1985, Gobas et al. 1986, Gobas and Mackay 1987). Based on these observations, a two-phase resistance model has been suggested, which assumes that gill uptake involves transport in aqueous and in lipid or membrane phases (Gobas and Mackay 1987). The resulting equations for k_1 and E_w are

$$1 / k_1 = \left(V_F / Q_W\right) + \left(V_F / Q_L\right) / K_{ow} \tag{15}$$

$$1 / E_W = \left(G_V / Q_W\right) + \left(G_V / Q_L\right) / K_{ow} \tag{16}$$

where Q_w and Q_L are transport parameters with units of l/day that represent the transport rates in the aqueous and the lipid phases of the fish. In essence, Equation 15 and Equation 16 demonstrate that uptake and elimination tend to be controlled by transport in the lipid phases if the chemical's K_{ow} is low. But with increasing K_{ow}, chemical transport in the aqueous phases of the fish becomes more important and ultimately dominates the kinetics, resulting in a constant k_1 and E_w with K_{ow}. This model satisfactorily describes the behavior of k_1 and E_w for most chemicals but it does not explain the drop of k_1 and E_w with K_{ow} for chemicals with extremely high K_{ow}. It is believed that the observed drop for high K_{ow} substances is not due to reduced gill uptake but the result of a reduced bioavailability and/or experimental errors associated with the difficult water concentration measurements (Gobas and Mackay 1987).

Equation 15 demonstrates that to estimate k_1, data are required for the chemical's K_{ow}, the weight of the fish V_F and the water and lipid phase transport parameters Q_w and Q_L. The following relationship between Q_w and V_F has been derived from experimental data and can be used to estimate Q_w from the weight of the fish (Gobas and Mackay 1987):

$$Q_w = 88 \cdot 3 \cdot V_F^{0.6(\pm 0.2)} \tag{17}$$

Because of insufficient data, a similar weight-dependent relationship for Q_L can not be derived at present. However, it appears from the available data that Q_L is approximately 100 times smaller than Q_w (Gobas and Mackay 1987). In particular for chemicals with a high bioconcentration potential (i.e., high K_{ow}), Q_L can therefore often be ignored and an accurate value is not needed.

Equations 15 and 17 provide a simple method to estimate k_1 from K_{ow} and the weight of the fish. For example, in a 0.25 kg fish, Q_w is 38.4 l/day and Q_L is approximately 0.384 l/day, thus giving a k_1 of 154 $l \cdot kg^{-1} \cdot d^{-1}$ for a chemical with a log K_{ow} of 6.

Gill Elimination

The rate at which chemicals are being eliminated by fish via the gills is expressed by the gill elimination rate constant k_2 which has units of 1/day. Models for the chemical elimination from the fish to the water via the gills are closely related to models for the chemical uptake rate constant since elimination is in essence the reverse process of gill uptake. The ratio of k_1 and k_2 can be viewed as the chemical's partition coefficient between the fish lipids and water, which can be approximated by $L_F \cdot K_{ow}$.

$$k_1 / k_2 = L_F \cdot K_{ow} \tag{18}$$

where

L_F = lipid content of the fish, i.e., the ratio of the lipid weight V_L (g) and weight V_F (g) of the fish, V_L/V_F. After substitution of Equation 15 into Equation 18, it follows that

$$1 / k_2 = \left(V_L / Q_W \right) \cdot K_{ow} + \left(V_L / Q_L \right) \tag{19}$$

where Q_w and Q_L are the same as in Equation 15 and Equation 16. Equation 19 predicts that k_2 is relatively constant if K_{ow} is low, and drops with increasing K_{ow} for chemicals of higher K_{ow}.

Metabolic Transformation

The rate at which chemicals are being metabolized in fish is expressed by the metabolic transformation rate constant k_M which has units of 1/day. Presently, there are few models that can be used to estimate k_M for organic substances in fish or other organisms. This is a serious knowledge gap, in particular when the purpose of the model is to estimate the bioaccumulation tendency of new chemicals for which no information regarding metabolic transformation exists.

However, if information regarding the metabolic transformation is available, an appropriate value for k_M can be estimated. For example, if the chemical's half-life $t_{1/2}$ is five years, a k_M of $0.693/t_{1/2}$ or 0.00038 d^{-1} can be estimated. This value may be small compared to k_2 or k_E, which makes its precise value irrelevant. So, if a chemical is known to be persistent, it is often possible to assume that k_M is zero, without affecting the model calculations significantly.

Dietary Uptake

The rate at which chemicals are absorbed by fish from the diet, i.e., via the gastrointestinal tract, is expressed by the dietary uptake rate constant k_D which has units of kg food/kg fish/day. The dietary uptake rate is the combined process of the food ingestion rate F_D (kg food/day) and the diffusion rate of the chemical across the intestinal wall and the faecal egestion rate F_E (Gobas et al. 1988). The extent to which chemical in the diet is actually absorbed by the organism can be expressed by the dietary uptake efficiency E_D, which is related to k_D by

$$k_D = E_D \cdot F_D / V_F \tag{20}$$

Although there is a considerable variability in the data, it has been shown that E_D is approximately 0.5 for chemicals with a log K_{ow} less than 6. With further increasing log K_{ow} (i.e., above 6), E_D shows a tendency to drop with increasing K_{ow} (Gobas et al. 1988). These observations can be explained by a two-phase resistance model for dietary uptake, which assumes that dietary uptake involves transport in aqueous and in lipid or membrane phases (Gobas et al. 1988). From this model it follows that:

$$1 / E_D = 5.3(\pm 1.5) \cdot 10^{-8} \cdot K_{ow} + 2.3(\pm 0.3) \tag{21}$$

Based on the work of Weininger (1978) a simple model based on fish bioenergetics can be used to estimate the feeding rate F_D as a function of temperature T and the fish's body weight V_F as follows:

$$F_D = 0.022 \cdot V_F^{0.85} \cdot \exp(0.06 \cdot T) \tag{22}$$

Equations 20 to 22 provide a simple method to estimate k_D from K_{ow}, the weight of the fish and temperature. For example, a 0.25 kg fish at 10°C has a feeding rate of approximately 12 g food/day or 0.012 kg food/day. If the food contains a chemical with a log K_{ow} of 6, E_D is approximately 42% and k_D is 0.42 · 0.012/ 0.25 or 0.021 kg food/kg fish/day.

Elimination by Fecal Egestion

The rate at which chemicals are being eliminated by egestion of fecal matter, i.e., via the gastrointestinal tract, is expressed by the fecal elimination rate constant k_E which has units of kg feces/kg fish/day. Presently, there are few data that can be used to estimate k_E. However, the available data suggest that the fecal egestion rate is approximately three to five times lower than the ingestion rate (Gobas et al. 1988, 1989a). We thus suggest that

$$k_E = 0.25 \cdot k_D \tag{23}$$

and assume that k_E is related to K_{ow} and the feeding rate in a similar manner as

k_D. The four times lower value of k_E is believed to be due to the effect that as a result of food digestion, the egestion rate is lower than the feeding rate. In addition, fecal matter may have a lower affinity for hydrophobic chemicals than the more organic rich food phase. The two effects combined cause k_E to be smaller than k_D, resulting in a concentration increase in the gastrointestinal tract (i.e., magnification). This increase in chemical concentration in the gastrointestinal tract due to food digestion provides the concentration gradient that is required for net uptake of the chemical into the fish, thus causing the concentration in the fish to exceed that in the consumed food (i.e., biomagnification) as long as gill elimination and metabolic transformation are small (Gobas et al. 1992). The ratio k_D/k_E is the predominant factor in the model that causes biomagnification in the food chain.

Growth

The rate of fish growth can have a significant effect on the steady-state concentration in the fish. Fish growth results in an increase of the fish volume and a drop in concentration if uptake is too slow to compensate the reduction in chemical mass per volume. The following generalized growth equations have been suggested by Thomann et al. (this volume) and are believed to give an adequate representation of the magnitude of fish growth:

$$k_G = 0.00251 \cdot V_F^{-0.2} \quad \text{for temperatures around } 25°C \quad (24)$$

$$k_G = 0.000502 \cdot V_F^{-0.2} \quad \text{for temperatures around } 10°C \quad (25)$$

FOOD-CHAIN ACCUMULATION

So far we have discussed chemical bioaccumulation in individual organisms. However, as a result of the feeding relationships, the chemical concentration in the predator is related to that in its prey. The trophodynamics thus play a role in the transfer of chemicals through the food chain and the accumulation of chemicals in the organisms of the food chain (Oliver and Niimi 1988, Thomann and Connolly 1984). In terms of describing and modeling this process we can combine the submodels for individual organisms to form food chains if feeding relationships can be defined. To include the effect of feeding interactions on the chemical concentration in the fish, we can add food preference factors P_i to Equation 13 to give

$$C_F = \left(k_1 \cdot C_{WD} + k_D \cdot \Sigma P_i \cdot C_{D,i}\right) / \left(k_2 + k_E + k_M + k_G\right) \quad (26)$$

where $\Sigma P_i \cdot C_{D,i}$ represents the composition of the fish's diet. P_i is the fraction of the fish's diet that consists of component i with concentration of $C_{D,i}$. If for example, the fish's diet consists of 20% phytoplankton, 70% benthic invertebrates and 10% small fish, then P_i is respectively 0.2, 0.7 and 0.1, adding up to 1, and $C_{D,i}$ is respectively C_A, C_B and C_F for small fish.

Since chemical uptake in phyto- and zooplankton is predominantly from the water, dietary accumulation (or biomagnification) can usually be ignored. Food consumption by phyto- and zooplankton may thus contribute insignificantly to the accumulation of chemicals in the food chain.

MODELING TOXICITY

Toxic effects resulting from chemical exposure can be expected to occur when the concentration of the chemical in the organism reaches a certain threshold level. If these threshold levels can be defined, it is possible to estimate toxic effects in biota from the chemical concentrations in the fish. For example, several studies have shown that for a large group of hydrophobic organic substances (often referred to as "narcotics") an internal concentration in the organism of approximately 1 to 3 mmol/kg causes acute lethality (McCarthy 1986, Abernethy and Mackay 1988, Van Hoogen and Opperhuizen 1988, Gobas et al. 1991). This concentration of approximately 2.0 mmol/kg appears to apply to several species of organisms including various species of fish, benthic invertebrates and plants, and to a number of hydrophobic inert substances such as alcohols, ketones, chlorinated aromatics, and alkanes. It is believed that this concentration represents a "minimal" or "basic" toxicity of a chemical substance. In other words, if a chemical does not act by a specific mode of toxic action, it exerts acute lethality at an internal concentration of approximately 2.0 mmol/kg. This suggests that many hydrophobic organic chemicals have a similar activity at the target site (i.e., a similar toxicity) and that the extent of the toxic impact is directly related to the concentration in the organism. Differences in observed LC_{50}s and "sensitivities" between various organisms are therefore considered to be a reflection of the rates of chemical uptake in the organisms, causing the threshold levels to be reached at different times in different organisms.

Because this mode of toxicity appears to be chemical independent, it is not surprising that several studies have demonstrated that the toxicity of mixtures of these simple hydrophobic organics is additive (Hermens 1989). Consequently, it can be suggested that acute lethality occurs if the sum of the chemical concentrations in the organisms reaches the threshold level. From a modeling perspective this means that the occurrence of acute lethality can be treated as the sum of the toxic contributions of the individual chemical components, i.e., $\Sigma X_i / V_F$, where X_i is the amount of each chemical i in the mixture (in mmol) and V_F is the volume of the organism (in kg). When $\Sigma X_i / V_F$ reaches the toxic threshold level(s), acute lethality is predicted.

An interesting feature of this model is that it is possible to determine which component of a mixture of chemicals in water or sediment exerts the greatest toxic impact. This is the chemical which causes the highest concentration in the organism (i.e., the nearest to the toxic threshold level). This information may be important when setting priorities for chemical clean up. Another convenient aspect of this modeling approach is that LC_{50} values for individual chemicals in various organisms are not needed to make predictions of chemical toxicity in a variety of organisms.

MODEL APPLICATION

To test the food chain model, we applied the model to the Lake Ontario food chain. This was possible thanks to the extensive collection of data on chemical concentrations in the most important species in lake Ontario (Oliver and Niimi 1988). We refer to these original studies for a complete account of the observed data. The model considers phytoplankton, zooplankton (i.e., *Mysis relicta*), two benthic invertebrate species, (i.e., Oligochaetes (*Tubifex tubifex*) and *Pontoporeia affinis*), and four fish species, i.e., sculpins (*Cottus cognatus*), alewifes (*Alosa pseudoharengus*), smelt (*Osmerus mordax*), and a composite group of 60 large size salmonid species, including Lake Trout (*Salvelinus namaycush*), Rainbow trout (*Salmo Gairdneri*), and Coho Salmon (*Oncorhynchus velinus namaycush*). To compare the model predictions to field observations, data regarding lipid contents and weights of the individual organisms were taken from the study of Oliver and Niimi (1988). Data for feeding preferences were taken from Flint (1986). Based on experimental observations that approximately 50% of the total concentration of PCBs in Lake Ontario is in a sorbed state (IJC 1991), a value of $2.5 \cdot 10^{-7}$ kg/l was chosen for the organic matter content of Lake Ontario water, representing a C_{WD} that is 50% of C_{WT} for a chemical with a log K_{ow} of 6.6. A summary of the data that were used in the model are listed in Table 2.

Typical results of the model-data comparison are graphically illustrated in Figure 1 for phyto- and zooplankton, in Figure 2 for benthic invertebrates and in Figure 3 for various species of fish. Each of these model-data comparisons consist of data for approximately 60 organic substances, including several PCB congeners, DDT, DDE, chlorobenzenes, mirex, octachlorostyrene, hexachlorobutadiene, lindane, and others. Typical results are also presented in Table 3, which summarizes predicted and observed concentrations of total PCBs in various organisms of the Lake Ontario food chain. Figure 4 further illustrates the good agreement with observed data. The only exception is that observed concentrations in phyto- and zooplankton are higher than predicted. The reason for this is unknown, but it is possible that sampling difficulties and small sample numbers (n = 3 for phytoplankton and n = 2 for zooplankton) contribute to the poor fit. The apparent underestimation of the observed concentrations in phyto- and zooplankton does not appear to affect the quality of fit for the fish species.

Table 2.   A Summary of the Input Parameters that are used in the Food Chain Model Calculations.

Water temperature: 8°C
Organic content of the water: 0.00000025 kg/l
Organic carbon content of the sediments: 2%
Density of lipids: 0.9 kg/l
Density of organic carbon: 0.9 kg/l
Metabolic transformation rate constant: 0 d^{-1}

Species characteristics:
Phytoplankton:
   Lipid content 0.5%

Zooplankton: *Mysids*
   Lipid content 5.0%

Benthos 1: *Pontoporeia*
   Lipid content 3.0%

Benthos 2: *Oligochaetes*
   Lipid content 1.0%

Fish 1: Sculpin
   Weight: 5.4 g; lipid content: 8.0%
   Diet: 18% zooplankton, 82% *Pontoporeia*

Fish 2: Alewife
   Weight: 32 g; lipid content: 7.0%
   Diet: 60% zooplankton, 40% *Pontoporeia*

Fish 3: Smelt
   Weight: 16 g; lipid content: 4.0%
   Diet: 54% zooplankton; 21% *Pontoporeia*; 25% sculpins

Fish 4: Salmonids
   Weight: 2150 g; lipid content: 16%
   Diet: 10% sculpin, 50% alewife, 40% smelt

This may be an indication that actually observed concentrations for phyto- and zooplankton may have exceeded the real values.

A sensitivity analysis showed that contaminant concentrations in all fish species are more sensitive to changes in concentration in the sediments than in

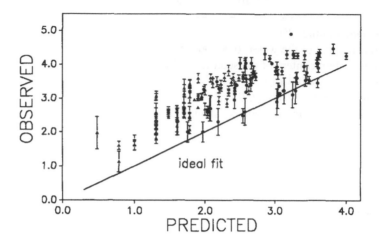

Figure 1. **Logarithms of observed concentrations (ng/kg) of various organochlorines in phytoplankton (▲) and in zooplankton (●)
versus model predicted concentrations. The solid line represents
the ideal fit.**

the water. This indicates that, in general, concentrations in Lake Ontario fish are
more responsive to sediment than to water concentrations.

CONCLUSIONS

This study presents a model for the chemical distribution in aquatic food
chains. The model consists of a series of equations for the chemical uptake,
elimination, and bioaccumulation in fish, aquatic macrophytes, benthos, and
zooplankton. Each of these "submodels" has been tested individually and
represents the current state of understanding regarding the mechanism of
chemical bioaccumulation. The food chain model combines these expressions
and the combined model is shown to be in good agreement with observed data
for the Lake Ontario food chain.

This ability to predict chemical distribution and food chain transfer of organic
substances in real food chains provides an important tool for the management of
contaminants on an "ecosystem" level. The model is simple since it only requires
basic data to characterize the food chain, such as organism weights, lipid

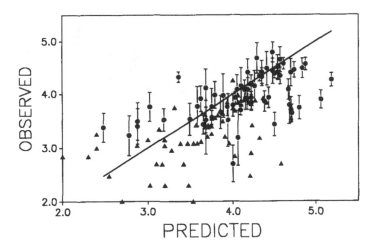

Figure 2.   **Logarithms of observed concentrations (ng/kg) of various organochlorines in Oligochaetes (▲) and in *Pontoporeia* (●) versus
model predicted concentrations. The solid line represents the
ideal fit.**

contents, and trophic interactions. Thus with little effort, predictions of chemical
concentrations in various organisms can be made from concentrations in water
and sediments. The model predictions are believed to be accurate within a factor
of two to three.

Recently, models for the physical distribution of chemicals have been
developed, which can successfully estimate contaminant concentrations in water
and sediments from chemical discharges (IJC 1991). The combination of these
fate models with food chain models will provide a capability to assess on an
"ecosystem" level the chemical exposure of biota as a result of chemical
discharges. Finally, it is possible to formulate models to estimate toxic impacts,
possibly following the threshold approach outlined above. This requires that
toxic effects are associated with a specific concentration in the organism.
Presently, such a threshold level has only been identified for the acute lethality
of narcotics. This model is difficult to field test since environmental concentrations are usually too low to reach the threshold levels required to cause the toxic
effects. Further research may reveal that other, more subtle toxic effects can be
related to a specific threshold concentration, thus providing an ability to interpret
chemical loadings in terms of toxic effects to the ecosystem.

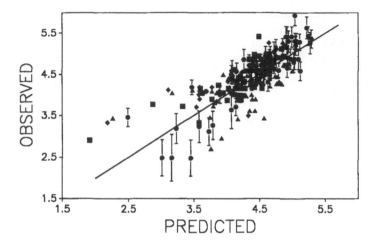

Figure 3. Logarithms of observed concentrations (ng/kg) of various organochlorines in Salmonids (●), smelt (◆), sculpins (▲) and
alewife (■) versus model predicted concentrations. The solid
line represents the ideal fit.

**Table 3. Observed and model predicted concentrations (μg/g) of total
PCBs in various organisms of the Lake Ontario food chain [a]**

Species	Predicted	Observed
Phytoplankton	0.011	0.05 (±0.012)
Mysids	0.11	0.33 (±0.12)
Photoporeia	0.86	0.79 (±0.48)
Oligochaetes	0.29	0.18 (±0.1)
Sculpins	1.6	1.6
Alewifes	0.99	1.3
Smelt	1.4	1.4
Salmonids	3.5	3.7 (±0.45)

[a] Observed data are from Oliver and Niimi (1988).

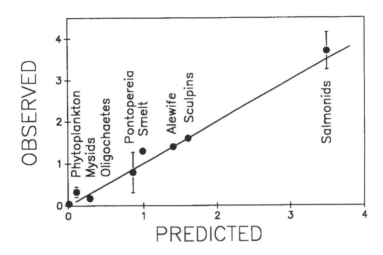

Figure 4. Observed concentrations (ng/kg) of total PCBs in organisms of the lake Ontario food chain versus predicted concentrations. The solid line represents the ideal fit.

REFERENCES

Abernethy, S.G., D. Mackay and L.S. McCarthy. "Volume Fraction Correlation for Narcosis in Aquatic Organisms: The Key Role of Partitioning," *Environ. Toxicol. Chem.* 7:469-481 (1988).

Black, M.C., and J.F. McCarthy. "Dissolved Organic Macromolecules Reduce the Uptake of Hydrophobic Organic Contaminants by the Gills of the Rainbow Trout (*Salmo Gairdneri*)," *Environ. Toxicol. Chem.* 7:593-600 (1988).

Bruggeman, W.A., L.B.J.M. Matron, D. Kooiman and O. Hutzinger. "Accumulation and Elimination Kinetics of Di- , Tri- , and Tetra Chlorobiphenyls by Goldfish after Dietary and Aqueous Exposure," *Chemosphere* 10:811-832 (1981).

Clayton, J.R., S.P. Pavlou and N.F. Breitner. "Polychlorinated Biphenyls in Coastal Marine Zooplankton," *Environ. Sci. Technol.* 18:676-682 (1977).

Clark, K.E., F.A.P.C. Gobas and D. Mackay. "Model of Organic Chemical Uptake and Clearance by Fish from Food and Water," *Environ. Sci. Technol.* 24:1203-1213 (1990).

DiToro, D.M. "A Particle Intercation Model of Reversible Organic Chemical Sorption," *Chemosphere* 15:1503-1538 (1985).

Flint, R.W. "Hypothesized Carbon Flow Through the Deepwater Lake Ontario Food Web," *J. Great Lakes Res.* 12:344-354 (1986).

Fox, M.E., J.H. Carey and B.G. Oliver. "Compartmental Distribution of Organochlorine Contaminants in the Niagara River and the Western Basin of Lake Ontario," *J. Great Lakes Res.* 9:287-294 (1983).

Geyer, H., G. Politzki and D. Freitag. "Prediction of Ecotoxicological Behaviour of Chemicals: Relationship Between n-Ocatnol-Water Partition Coefficient and Bioaccumulation of Organic Chemicals by Alga Chlorella," *Chemosphere.* 13:269-284 (1984).

Gobas, F.A.P.C., A. Opperhuizen and O. Hutzinger. "Bioconcentration of Hydrophobic Chemicals in Fish: Relationship with Membrane Permeation," *Environ. Toxicol. Chem.* 5:637-646 (1986).

Gobas, F.A.P.C., and D. Mackay. "Dynamics of Hydrophobic Organic Chemical Bioconcentration in Fish," *Environ. Toxicol. Chem.* 6:495-504 (1987).

Gobas, F.A.P.C., D.C.G. Muir and D. Mackay. "Dynamics of Dietary Bioaccumulation and Faecal Elimination of Hydrophobic Organic Chemicals in Fish," *Chemosphere* 17:943-962 (1988).

Gobas, F.A.P.C., K.E. Clark, W.Y. Shiu and D. Mackay. "Bioconcentration of Polybrominated Benzenes and Biphenyls and Related Superhydrophobic Chemicals in Fish: Role of Bioavailability and Faecal Elimination," *Environ. Toxicol. Chem.* 8:231-247 (1989a).

Gobas, F.A.P.C., D.C. Bedard, J.J.H. Ciborowski and G.D. Haffner. "Bioaccumulation of Chlorinated Hydrocarbons by the Mayfly *Hexagenia Limbata* in Lake St. Clair," *J. Great Lakes Res.* 15:581-588 (1989b).

Gobas, F.A.P.C., E.J. McNeil, L. Lovett-Doust and G.D. Haffner. "Bioconcentration of Chlorinated Aromatic Hydrocarbons in Aquatic Macrophytes (*Myriophyllum Spicatum*)," *Environ. Sci. Technol.* 25:924-929 (1991).

Gobas, F.A.P.C., E.J. McNeil, L. Lovett-Doust and G.D. Haffner. "A Comparative Study of the Bioconcentration and Toxicity of Chlorinated Hydrocarbons in Aquatic Macrophytes and Fish," *Am. Soc. Test. Mat.* 1115:174-183 (1991).

Gobas, F.A.P.C., J.R. McCorquodale and G.D. Haffner. "Intestinal Absorption and Biomagnification of Organochlorines," *Environ. Toxicol. Chem.* in press (1992).

Hermens, J.L.M. "Quantitative Structure-Activity Relationships of Environmental Pollutants," in *Handbook of Environmental Chemistry, Vol. 2E, Reactions and Processes*, O. Hutzinger, Ed. (Berlin: Springer-Verlag, 1989), pp.111-162.

International Joint Commission. "Proceedings of the Great Lakes Mass Balance Workshop," Niagara on the Lake, in press (1991).

Karickhoff, S.W. "Organic Pollutant Sorption in Aquatic Systems," *J. Hydraul. Eng. ASCE.* 110:707-735 (1984).

Landrum, P.F., M.D. Reinhold, S.R. Nihart and B.J. Eadie. "Predicting the Bioavailability of Organic Xenobiotics to Pontoporeia Hoyi in the Presence of Humic and Fulvic Materials and Natural Dissolved Organic Carbon," *Environ. Toxicol. Chem.* 4:459-467 (1985).

Landrum, P.F., T.D. Fontaine, W.R. Faust, B.F. Eadie and G.A. Lang. "Modelling the Accumulation of Polycyclic Hydrocarbons by the Amphipod Diporeia," This volume (1992).

Mallhot, H. "Prediction of Algal Bioaccumulation and Uptake Rate of Nine Organic Compounds by Ten Physiocochemical Properties," *Environ. Sci. Technol.* 21:1009-1013 (1987).

McCarthy, J.F., and B.D. Jimenez. "Reduction in Bioavailability to Bluegills of Polycyclic Aromatic Hydrocarbons Bound to Dissolved Humic Material," *Environ. Toxicol. Chem.* 4:511-521 (1985).

McCarthy, L.S. "The Relationship Between Aquatic Toxicity QSARs and Bioconcentration for some Organic Chemicals," *Environ. Toxicol. Chem.* 5:1071-1080 (1986).

McKim, J.M., P.K. Schnieder and G. Veith. "Absorption Dynamics of Organic Chemical Transport Across Trout Gills as Related to Octanol-Water Partition Coefficient," *Toxicol. Appl. Pharmacol.* 77:1-10 (1985).

Oliver, B.G., and A.J. Niimi. "Trophodynamic Analysis of Polychlorinated Bipehnyl Congeners and Other Chlorinated Hydrocarbons in the Lake Ontario Ecosystem," *Environ. Sci. Technol.* 22:388-397 (1988).

Shea, D. "Developing National Sediment Quality Criteria," *Environ. Sci. Technol.* 22:1256-1261 (1988).

Thomann, R.V. "Bioaccumulation Model of Organic Chemical Distribution in Aquatic Food Chains," *Environ. Sci. Technol.* 23:699-707 (1989).

Thomann, R.V., and J.P. Connolly. "Model of PCB in the Lake Michigan Lake Trout Food Chain," *Environ. Sci. Technol.* 18:65-71 (1984).

Thomann, R.V., J.P. Connolly and T. Parkerton. "Modelling Accumulation of Organic Chemicals in Aquatic Food-Webs," in Chapter 7 of this book (1992).

Van Hoogen, G., and A. Opperhuizen. "Toxicokinetics of Chlorobenzenes in Fish," *Environ. Toxicol. Chem.* 7:213-219 (1988).

Weininger, D. "Accumulation of PCBs by Lake Trout in Lake Michigan," Ph.D. thesis, University of Wisconsin-Madison (1978).

7 MODELING ACCUMULATION OF ORGANIC CHEMICALS IN AQUATIC FOOD WEBS

INTRODUCTION

The purpose of this paper is to provide an overview of models of organic chemical uptake and transfer in aquatic food webs. The degree to which the aquatic food chain may accumulate chemicals above some equilibrium value associated with uptake from the water has been of concern for a number of years. Models which consider both aqueous and dietary exposure routes can assist in determining the degree of bioaccumulation or the lack thereof. Furthermore, chemical food chain models (food chain and food web are used interchangably throughout this paper) in site-specific cases can provide valuable frameworks for determining the efficacy of environmental control actions in reaching target chemical concentrations in key aquatic organisms, such as fish species used in human diets.

Two general classes of models are described: (1) generic equilibrium models which are useful for screening analyses and provide insight into the principal mechanisms of chemical transport and accumulation, and (2) site-specific models which may be time- and age-dependent. In both cases, the aquatic system is divided into ecological compartments (e.g., phytoplankton, small fish, or a top predator in a specific age class) and abiotic compartments (e.g., water column and sediment). Figure 1 shows a generic compartment model with the notation to be used in this paper. Figure 2 shows a typical site-specific age-dependent food web for the striped bass in the Hudson River.

THEORY
Lipid and Organic Carbon Normalizations

Partitioning of organic chemicals into aquatic organisms is governed to first-order by the lipid pool of the organism (Mackay 1982, Connolly and Pedersen

ISBN 0-87371-511-X

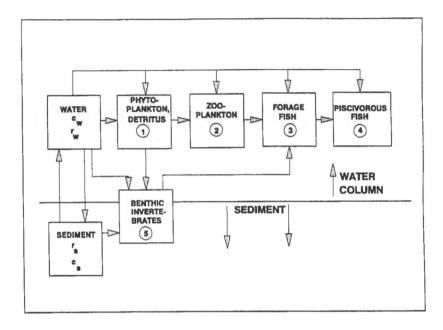

Figure 1. Schematic of a five-compartment generic food web model.

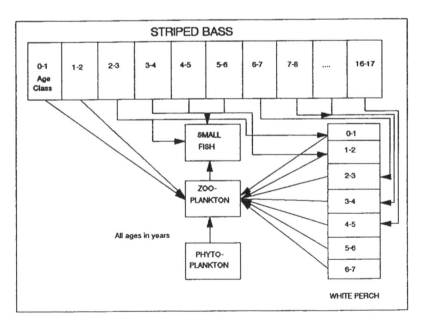

Figure 2. Compartments of age-dependent striped bass model for Hudson River (Thomann et al. 1989).

1988, Thomann 1989). Also, as noted in a review of sediment water quality criteria (USEPA 1989),and as discussed by Bierman (1990), the partitioning of organic chemicals is determined to a large degree by the amount of organic carbon present in the particulate matter. It is recognized that other components (e.g., protein) may influence the distribution of the chemical. Hallam et al. (1989) provide a detailed model of chemical uptake which includes this component. However, it is not expected that the distribution of chemical in these other components will be high due to the relatively low chemical affinity for such phases. The model equations presented herein are therefore written in terms of chemical concentrations in aquatic organisms on a lipid basis and for abiotic particles on an organic carbon basis.

The tendency for organic chemicals to partition into lipid and organic carbon pools is broadly represented by the octanol-water partition coefficient (K_{ow}). In this work, to first approximation, the preference for chemicals to partition to octanol, lipid, and organic carbon is considered identical.

The chemical concentration in an organism on a lipid basis, v[μg chemical/g(lipid)], is related to the wet weight chemical concentration, v_{wt}[μg chemical/g(wet)], and the fraction lipid, f_L[g(lipid)/g(wet)], by

$$v = \frac{v_{wt}}{f_L} \tag{1}$$

A plot of wet weight concentration against lipid fraction should therefore be linearly related. Figure 3 shows the relationship between the wet weight concentration of two PCB mixtures (Aroclors 1016 and 1254) for white perch for the Hudson River in the vicinity of Troy, New York. The correlation of the wet weight concentration to lipid fraction for the 1016 mixture, and to a lesser degree for the 1254 mixture, is apparent and thus the variability is significantly reduced by a normalization to lipid content of the fish. Figure 4 computed from the data in Gruger et al. (1975) shows the effect of lipid normalization to body components of the coho salmon for three PCB congeners. On a wet weight basis, the range across the components is almost an order of magnitude while on a lipid basis, the variability is reduced considerably. It is interesting to note that the brain PCB concentration is consistently lower than the other fish components. Figure 5 illustrates the importance of lipid normalizing in interpreting trophic transfers of chemicals. In the upper panel, the trend of total PCB on a wet weight basis is not clear with increasing trophic level. The lower panel however shows that when the data are lipid normalized, food chain accumulation appears to be occurring up through the striped bass as a top predator with a low lipid fraction.

Normalizations with organic carbon for abiotic compartments show similar reductions in the variability of chemicals sorbed to particulate matter expressed on a dry wet basis (e.g., Karickhoff et al. 1979, Karickhoff 1984, and USEPA 1989.)

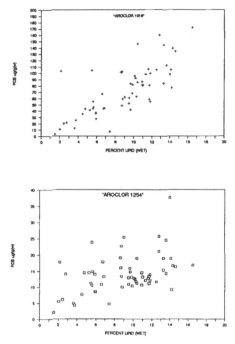

Figure 3. Wet weight PCB concentrations vs percent lipid for two Aroclors, white perch, Hudson River in vicinity of Troy New York. Compiled from data of NYS DEC.

General Equations

For any ecological compartment and/or age class within a compartment (see Figures 1 and 2), the mechanisms included in the general modeling framework are

1. Direct uptake of the dissolved chemical from the water by diffusive exchange across organism membranes
2. Food web accumulation of the chemical resulting from consumption of contaminated prey
3. Depuration of the chemical due to all loss pathways
4. Growth and respiration of the organism and the effect of such physiological factors on chemical concentration and the rates of uptake of chemical
5. Migration of the organism through temporally and spatially varying water concentrations

Uptake from Water Only

The uptake and subsequent internal transport of chemicals from water by aquatic organisms has been the subject of a number of studies. See, for example,

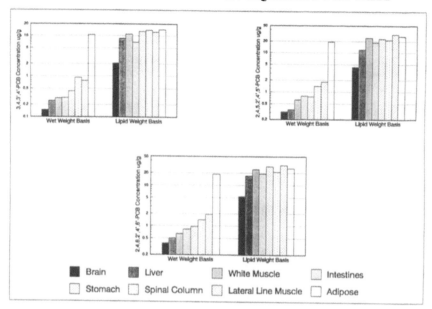

Figure 4.   Comparison of PCB concentrations in compartments of juvenile coho salmon on a wet weight and lipid normalized basis. Compiled from data in Gruger et al. 1975.

Norstrom et al. (1976), Gobas et al. (1986), Gobas and Mackay (1987), Barber et al. (1988), and Erickson and McKim (1990).

Assuming an organism is exposed to a chemical in the water phase only, then a mass balance equation around the average organism in a compartment may be written as an exchange process across a lipoprotein membrane as follows:

$$\frac{dV_m}{dt} = k_u w_L \left(c_{w,s} - c_B \right) - K_1 V_m \tag{2}$$

where V_m is the chemical whole body burden [μg chemical/organism], k_u is the chemical uptake rate [L/day-kg(lipid)], w_L is the lipid weight, $c_{w,s}$ is the "freely available" dissolved chemical in the water column or in the interstitial water of the sediment (see Figure 1), c_B is the "free" concentration of the chemical in the blood and K_1 is the loss rate of chemical due to mechanisms other than reverse transfer across the membrane (e.g., losses from skin surface, fecal losses, and chemical metabolism). If the blood concentration is considered in equilibrium with the tissue concentration v(μg/kg/(lipid)) then

$$\frac{v}{c_B} = N_{wl} \tag{3}$$

where N_{wl} is a lipid concentration to blood concentration partition coefficient. To first approximation,

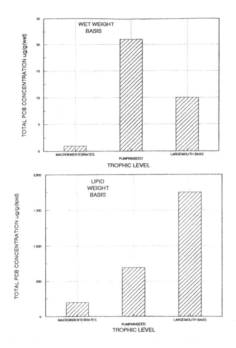

**Figure 5. Comparison of PCB concentrations in a Hudson River food
chain. Upper: on wet weight basis; lower: on lipid basis. Compiled from data of NYS DEC.**

$$N_{wl} \approx K_{ow} \qquad (4)$$

and Equation 2 is

$$\frac{dv_m}{dt} = k_u w_L \left(c_{w,s} - \frac{v}{K_{ow}} \right) - K_1 v_m \qquad (5)$$

Since the whole body burden is related to the chemical concentration by

$$v_m = v w_L \qquad (6)$$

the chemical concentration on a lipid basis is given by

$$\frac{dv}{dt} = k_u c_{w,s} - K' v \qquad (7)$$

where

$$K' = \frac{k_u}{K_{ow}} + K_1 + G \tag{8}$$

for $G(\text{day}^{-1})$ as the growth rate of organism lipid.

For zero K_1 and growth, Equation 8 shows that the excretion rate is inversely related to the K_{ow} of the chemical.

At equilibrium, a lipid-based whole body Bioconcentration Factor (BCF) [μg/kg(lipid) $\div$ μg/L(water)] can be defined from Equation 7 as

$$N_w = \frac{v}{c_{w,s}} = \frac{k_u}{K'} \tag{9}$$

The BCF is therefore the ratio of the chemical concentration (from water exposure only) to the freely dissolved chemical concentration, either in the water column or in the sediment interstitial water.

Uptake from Water and Food

In addition to the chemical intake from water, the organism also receives chemical input from consumption of contaminated food. This mass input of chemical depends on the rate of feeding, the chemical concentration of the food source and the assimilation efficiency of the chemical.

The general mass balance equation for the whole body burden for a given compartment, i, is then similar to Equation 2 for water uptake but with the additional mass input due to feeding (Norstrom et al. 1976, Thomann 1981, Thomann and Connolly 1984, Connolly and Tonelli 1985, Thomann 1989, Connolly 1991). Therefore,

$$\frac{dv_i}{dt} = k_{ui}c_{w,s} - K_i'v_i + \Sigma_j \alpha_{ij}p_{ij}I_{L,i}v_j \tag{10}$$

where

$i = 2...n$

for "n" compartments above the phytoplankton and where α_{ij} is the chemical assimilation efficiency (g chemical absorbed/g chemical ingested); p_{ij} is the food preference of i on j; $I_{L,i}$ is the lipid-specific consumption of organism i, (g(lipid) prey/g(lipid) predator-d); and t is real time (days).

For a simple chain where organism i feeds only on i-1,

$$\frac{dv_i}{dt} = k_{ui}c_{w,s} + \alpha I_{L,i.i-1}v_{i-1} - K_i'v_i \tag{11}$$

where

$$I_{L,i.i-1} = P_{i,i-1}I_{L,i} \tag{11a}$$

At equilibrium,

$$v_i = \frac{k_{ui}c_{w,s}}{K_i'} + g_{i,i-1}v_{i-1} \tag{12}$$

for $g_{i,i-1}$ as a food chain multiplier given by

$$g_{i,i-1} = \frac{\alpha I_{L,i,i-1}}{K_i'} \tag{12a}$$

The extent of any food chain magnification above that for exposure to water only is then given by the interaction of net chemical intake ($\alpha \cdot I$) and overall loss rate, K'. Chemicals that are excreted slowly and assimilated strongly then will tend to accumulate more readily in the food chain.

For a pelagic system, the phytoplankton are considered as the base of the food chain (compartment #1, see Figure 1). The relationship between the chemical concentration in the phytoplankton and that in the water over a range of K_{ow} is a subject of some research. Field and laboratory data (e.g., Lederman and Rhee 1982, Oliver and Niimi 1988, see also Connolly 1991 for further review and discussion of data) indicate a phytoplankton BCF approximately equal to log K_{ow} up to about 5–6. Thereafter, the data indicate a constant BCF independent of K_{ow}. Swackhammer (unpublished data) has shown that the BCF for the phytoplankton appears to be related to the growth stage of the phytoplankton biomass.

The mass of chemical in the phytoplankton per volume of water $v_1[\mu g/L]$ is related to the phytoplankton chemical concentration $v_1[\mu g/kg(lipid)]$ by

$$v_1 = v_1 w_1 f_{L1} \tag{13}$$

where w_1 [kg(w)/L] is the total phytoplankton biomass and f_{L1} [kg(lipid)/kg(wet)] is the fraction lipid of the phytoplankton.

A mass balance equation for the phytoplankton chemical mass per volume of water using sorption-desorption kinetics is then given as

$$\frac{dv_1}{dt} = k_{u1}w_1c_w - K_1v_1 \tag{14}$$

Using Equation 13 gives the equation for the concentration as

$$\frac{dv_1}{dt} = k_{u1}c_w - K_1v_1 - G_1v_1 \tag{15}$$

where G_1 is the net growth rate of the total phytoplankton biomass. The phytoplankton BCF, N_1, at equilibrium is then given by

$$N_1 = \frac{v_1}{c_w} = \frac{k_{u1}}{K_1 + G_1} \tag{16}$$

The effect of the phytoplankton biomass growth rate may then influence the phytoplankton BCF at higher K_{ow} levels where the excretion rate is presumably low.

Parameter Estimation

The preceding model framework contains two broad classes of parameters: those associated with organism physiology (e.g., growth and respiration rates) and those associated with the specific chemical (i.e., uptake and depuration rates and chemical assimilation efficiency.)

Chemical Parameters
Uptake Rate

The chemical uptake rate is related to the respiration rate of the organism and the efficiency of transfer of the chemical across the organism membrane. One expression is given by (see e.g., Connolly 1991)

$$k_{ui} = \frac{a_{oc}a_c\rho}{a_{wd}f_Lc_{o2}}\beta \tag{17}$$

where a_{oc} is oxygen to carbon ratio, a_c is the carbon to dry weight ratio, a_{wd} as the wet to dry ratio, β is the ratio of chemical transfer efficiency (E_c) to oxygen transfer efficiency (E_o), ρ is the organism oxygen respiration rate [g(w)/g(w)-

day] and Co_2 is the oxygen concentration (mg/l). E_c has been shown to be a function of K_{ow} e.g., McKim (1985) with high K_{ow} (less than six) showing a decline in E_c. Thomann (1989) has expressed $E_c(K_{ow})$ as

$$For \log K_{ow} = 2\text{--}3: \quad Log E_c = -1.5 + 0.4 \log K_{ow}; for\ w > 10\text{--}100g(w)$$

$$For \log K_{ow} = 2\text{--}5: \quad Log E_c = -2.6 + 0.5 \log K_{ow}; for\ w < 10\text{--}100g(w)$$

$$For \log K_{ow} = 3\text{--}6: \quad\quad E_c = 0.5; \quad\quad\quad\quad\quad for\ w > 1\ 0\text{--}100g(w) \quad (18)$$

$$For \log K_{ow} = 5\text{--}6: \quad\quad E_c = 0.8; \quad\quad\quad\quad\quad for\ w < 10\text{--}100g(w)$$

$$For \log K_{ow} = 6\text{--}10: \quad Log E_c = 1.2 - 0.25 \log K_{ow}; \quad for\ w > 10\text{--}100g(w)$$

$$Log E_c = 2.9 - 0.5 \log K_{ow}; \quad for\ w < 10\text{--}100g(w)$$

Excretion Rate

Equation 8 indicates

$$K = \frac{k_u \left(E_c \left(K_{ow} \right), \rho \right)}{K_{ow}} + K_1 + G \quad\quad (19)$$

Equation 19, for $K_1 \ll G$, has been shown by Thomann (1989) to be approximately representative of observed excretion data for fish.

Chemical Assimilation Efficiency

The chemical assimilation efficiency is also an apparent function of K_{ow}. A summary of available data for fish (using statistical routines in Wilkinson 1988) is shown in Figure 6 (Parkerton, unpublished data). The scatter in the data is large but there is a clear decline in α for log K_{ow} greater than about 6.5. In the sediment food web model discussed below, α and E_c/E_0 were assigned using Figure 6 as follows: for log $K_{ow} = 2\text{--}4$, Equation 18 (for $w < 10\text{--}100g(w)$) was used, for log $K_{ow} = 4.5\text{--}6.5$, α and $E_c/E_0 = 0.7$; for log $K_{ow} = 7.0, 7.5, 8.0, 8.5,$ and 9.0, α and $E_c/E_0 = 0.5, 0.3, 0.1, 0.05,$ and 0.01, respectively.

Organism Parameters

The above equations for chemical accumulation are related to the bioenergetics of the organism via growth, feeding and respiration rates. Norstrom et al. (1976), Hallam et al. (1989) and Connolly (1991) discuss organism energetics for use in chemical accumulation models.

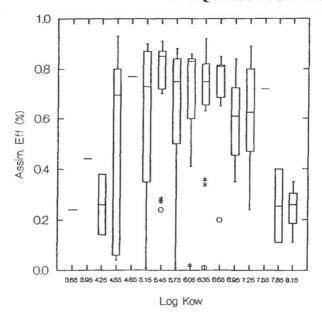

**Figure 6.   Variation of chemical asimilation efficiency for fish. Legend:
centerline of bar = median, upper and lower limits of bar = 25th
& 75th percentiles, upper and lower lines of bar = 5th and 95th
percentiles, asterisks, circles and dashes = outliers.**

The energy usage rate P_i[cal/g(w)–d] as given by Connolly (1991) is

$$P_i = \lambda_i \left(\rho_i + G_i\right) \tag{20}$$

for growth rate, G and respiration rate, ρ in [g(w)/g(w)-d] and λ_i in [cal/g(w)$_i$]. The energy intake rate by the animal is then the energy usage rate divided by the fraction of ingested energy that is assimilated, a. The food consumption rate, C_i in [g(w)$_i$/g(w)$_{i-1}$-d] is then given by

$$I_i = \frac{\lambda_i}{\lambda_{i-1}} \frac{\rho_i + G_i}{a} \tag{21}$$

for λ_{i-1} as the caloric density [cal/g(w)$_{i-1}$] of the food. Connolly (1991) assumes differences in caloric density to be related to the wet weight to dry weight ratio, i.e., the caloric density of dry tissue was assumed to be the same for predator and prey. Therefore, the lipid-specific consumption rate for use in Equation 10 is given by

$$I_{L,i} = \frac{(G+\rho)}{a}\left(\frac{a_{wd,i-1}}{a_{wd,i}}\right)\left(\frac{f_{L,i}}{f_{L,i-1}}\right) \tag{22}$$

where $a_{wd,i}$ and $a_{wd,i-1}$ are the wet-dry weight ratios for predator and prey, respectively. For organisms feeding on sediment organic carbon,

$$I_{LOC,i} = \frac{G+\rho}{a \cdot a_{wd,i}}\left(\frac{f_{oc,i}}{f_{L,i}}\right) \tag{22a}$$

where $I_{LOC,i}$ is in g(OC)sed/g(lipid)$_i$-day and $f_{oc,i}$ is the fraction organic carbon of the predator.

Growth Rate

The growth rate of an organism may be estimated from the weight-age data according to

$$G = \frac{\left(dw / dt_a\right)}{w} \tag{23}$$

for t_a = age.

A generalized growth rate expression is given by (see summary in Thomann 1981)

$$G(T) \approx 0.00586(1.113)^{T-20} w^{-0.2} \tag{24}$$

for T in °C.

Respiration Rate

The generalized respiration weight relationships for two temperatures as given in Norstrom (1976), (see also Thomann 1981) provide a means for estimating the respiration temperature relationships for organisms in the lower levels of the food chain. These relationships can also be used in generic chemical models.

$$\rho(T) \approx 0.0263(1.065)^{T-20} w^{-0.2} \qquad (25)$$

The general form for the respiration rate of fish in site specific dynamic computations follows that used by Thomann and Connolly (1983) and Connolly and Tonelli (1984). Thus, the respiration rate for those conditions is given by

$$\rho = bw^{\gamma} e^{dT} e^{vU} \qquad (26)$$

for U, the swimming speed given by

$$U = \omega w^{\delta} e^{\phi T} \qquad (27)$$

where T is the water temperature (°C) and $b, \gamma, d, v, \omega, \delta$ and ϕ are coefficients set at 0.043, 0.3, 0.03, 0.0176, 1.19, 0.32 and 0.045, respectively, based on the aforementioned previous work.

EQUILIBRIUM GENERIC MODELS
Pelagic Models

A simple four level generic food chain can be used to calculate the expected concentration in a compartment such as a piscivorous fish. The approach follows that of Thomann (1981, 1989). Assume that in Figure 1 there is no interaction with the sediment.

Let N_i be the Bioaccumulation Factor (BAF) for the ith level on a lipid basis ($\mu g/kg(lipid) \div \mu g/L$) as

$$N_i = \frac{v_i}{c_w} \qquad (28)$$

Then the equilibrium BAF for each level above the phytoplankton is given from Equation 12 as

$$N_2 = N_{2w} + g_{21}N_1 \qquad (28a)$$

$$N_3 = N_{3w} + g_{32}N_{2w} + g_{32}g_{31}N_1 \qquad (28b)$$

$$N_4 = N_{4w} + g_{43}N_{3w} + g_{43}g_{32}N_{2w} + g_{43}g_{32}g_{21}N_1 \qquad (28c)$$

where N_{iw} is the BCF for Equation 9. Neglecting growth and non-excretory loss mechanisms, $N_{iw} \approx K_{ow}$ and then

$$N_2 / K_{ow} = 1 + g_{21} \tag{29a}$$

$$N_3 / K_{ow} = 1 + g_{32} + g_{32}g_{21} \tag{29b}$$

$$N_4 / K_{ow} = 1 + g_{43} + g_{43}g_{32} + g_{43}g_{32}g_{21} \tag{29c}$$

The accumulation of the chemical above an equilibrium with K_{ow} (an increase in the fugacity or chemical potential as discussed by Connolly and Pedersen 1988), is therefore seen to increase with trophic level.

The behavior of $g_{i,i-1}$ can be obtained to first approximation by considering the allometric relations of respiration and growth from Equations 24 and 25. Using Equation 17 with $a_{oc} = 2.67$, $a_{wd} \approx 5$ to 6, $a_c = 0.4$, $E_o = 0.8$, and $c_{02} = 8.5$ mg/l, the uptake rate is given approximately by

$$k_u \approx 1000 \frac{w^{-0.2}E_c}{f_L} \tag{30}$$

Substitution of this equation and Equation 19 into Equation 12a and using Equation 22 with approximately constant lipid and wet-dry weight ratios gives

$$g \approx \frac{0.046\alpha(K_{ow})}{a\left(\dfrac{1000E_c}{f_L K_{ow}} + 0.01\right)} \tag{31}$$

As seen, g is approximately independent of the weight of the organism. Also, it is interesting to note that for $K_{ow} > 10^6$, $f_L \approx 0.10$ and $E_c = 0.1$ to 1.0,

$$g \approx (2 \rightarrow 5)\frac{\alpha(K_{ow})}{a} \tag{31a}$$

indicating that for this K_{ow} region, the food chain multiplier is proportional to the ratio of the chemical uptake efficiency and the food assimilation efficiency.

If Equation 18 is used for both E_c and for α, then

$$g \cdot a \approx \frac{0.046}{\dfrac{1000}{f_L K_{ow}} + \dfrac{0.01}{\alpha(K_{ow})}} \tag{32}$$

Figure 7 shows a plot of (g·a) versus K_{ow} using Equation 18 for α (K_{ow}). As seen in this figure, (g·a) is small for log K_{ow} less than about 4 to 5, increases to

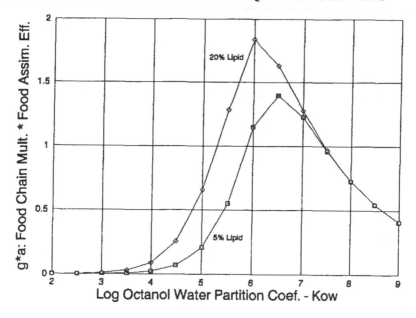

Figure 7. Variation of the product of food chain multiplier and food assimilation efficiency with octanol-water partition coefficient using Equation 32 and Equation 18 for w > 10 to 100 g(w).

a peak at log K_{ow} of about 6 to 6.5 and then declines. Thus, food chain effects are generally not significant for chemicals with log K_{ow} less than about 4 to 5. For more lipophilic chemicals, i.e., log K_{ow} greater than five and less than seven, food chain effects may be significant. For example, for log K_{ow} of six, and a food chain of organisms of about 10% lipid, the $(g \cdot a)$ factor is about 1.2. If the average food assimilation efficiency is 0.8, then g is 1.5. Using Equation 29c, the top predator would then have a BAF/K_{ow} of about eight times that due to uptake from the water only.

Sediment-Pelagic Models

A generic modeling framework for the accumulation of chemicals in aquatic systems which includes interaction with sediment chemical and sediment biota can also be derived from the previous equations (Thomann et al. 1992). Concern has been expressed for sediment-mediated transfer of chemicals in producing fish lesions (e.g., Malins et al. 1984) and in possible trophic transfer (Connor 1984). The bioavailability of chemicals from sediments is also of concern in establishing sediment quality criteria (USEPA 1989). A large number of laboratory experiments of chemical uptake from sediments by benthic invetebrates and to a lesser degree by fish have been conducted (e.g., Fowler et al. 1978, Oliver 1987, Landrum et al. 1989, Rubenstein et al. 1984). Likewise, field relationships between sediment chemical concentration and aquatic biota have been measured by many investigators (e.g., Oliver 1987, Oliver and Niimi 1988, Nalepa and Landrum 1988, Mudroch et al. 1989, Pereira et al. 1988, Huckins et al. 1988).

Bierman (1990) has summarized a considerable data base relating sediment concentrations to concentrations in benthic invertebrates and fish. In that work, it was concluded that clear relationships were not evident between organic carbon-normalized sediment and lipid-normalized biota. The analysis of Bierman (1990) is, however, based on a simple partioning between sediment and biota.

Model Structure and Equations

The compartmental structure of the food web examined in this model is shown in Figure 1. Five interactive biological compartments are considered together with the particulate and dissolved components in the water column and sediment. Benthic invertebrates obtain chemical via uptake from a combination of interstitial water and overlying water and direct ingestion of chemical on sediment particles and/or from phytoplankton and detrital material at the sediment-water interface. Forage fish accumulate chemical directly from the overlying water and from food in some linear combination from zooplankton and benthic invertebrates. By allowing the benthic community to interact with both the sediment and the overlying water column, the chemical transfer between the sediment and the overlying water must also be included. The equation for each of the biological compartments is developed in turn.

Applying Equation 10, the mass balance equation for the chemical in the benthic compartment (#5) is given at steady-state by

$$\frac{dv_5}{dt} = 0 = \left[k_{u5}\left(b_{5s}c_s + b_{5w}c_w\right)\right] + \left[\left(p_{5s}\alpha_{5s}I_{loc.5}\right)r_s + \left(p_{51}\alpha_{51}I_{L,5}\right)v_1\right]$$
$$-\left[\left(K_5'\right)v_5\right] \tag{33}$$

The first bracketed term on the right hand side of this equation represents the uptake of available chemical by benthic organisms from the sediment interstitial water and the overlying water column where b_{5s} and b_{5w} are the fraction of uptake from sediment and overlying water, respectively, $(b_{5s} + b_{5w} = 1)$. The second bracketed term represents the uptake of chemical from ingestion of sediment (r_s; µg/g (org C)) and phytoplankton (v_1) where p_{5s} and p_{51} are the preference for sediment and phytoplankton, respectively $(p_{5s} + p_{51} = 1)$. The third term is the loss of chemical due to excretion (K) and growth (G).

For the third compartment, the forage fish, the model equation is given by

$$\frac{dv_3}{dt} = 0 = k_{u3}c_w + p_{32}\alpha_{32}I_{L,3}v_2 + p_{35}\alpha_{35}I_{L,3}v_5 - \left(K_3 + G_3\right)v_3 \tag{34}$$

In this equation, p_{35} and p_{32} represent the relative feeding preference of forage fish for benthic invertebrates and zoooplankton, respectively, ($p_{35} + p_{32} = 1$). Note that if $p_{35} = 1.0$, then this compartment represents a bottom dwelling fish feeding only on the benthic community.

The phytoplankton, zooplankton and piscivorous fish equations are given by application of Equation 10.

In order to assess relative bioaccumulation of chemical from the sediment or the overlying water column, the following ratios can be defined.

Let v_i/r_s be the ratio of the organism chemical concentration on a lipid basis to the sediment chemical concentration on a carbon basis (with units µg/g(lipid) ÷ µg/g(organic carbon) or g(organic C)/g(lipid)). This ratio has been termed the Biota Sediment Factor (BSF), by Parkerton (1991). Also, let

$$\pi_s = \frac{r_s}{c_s} \tag{35}$$

$$\pi_{ws} = \frac{r_s}{c_w} \tag{36}$$

Equation 35 is the sediment partition coefficient, i.e., the ratio of the sediment chemical concentration (organic carbon basis) to the interstitial freely dissolved chemical concentration. Equation 36 is the partitioning between the sediment chemical concentration and the overlying water freely dissolved concentration.

The BSF is given from Equation 33 as

$$S_5 = \frac{v_5}{r_s} = \frac{N_{5w}}{\pi'} + g_{5s} + g_{51} \frac{N_{1w}}{\pi_{ws}} \tag{37}$$

where

$$\pi' = \frac{\pi_s \pi_{ws}}{b_{5s} \pi_{ws} + b_{5w} \pi_s} \tag{37a}$$

$$g_{5s} = \frac{p_{5s} \alpha_{5s} I_5}{K_5 + G_5} \tag{37b}$$

$$g_{51} = \frac{p_{51} \alpha_{51} I_5}{K_5 + G_5} \tag{37c}$$

The first term of Equation 37 represents the uptake of chemical from the water phase, either sediment interstitial water or overlying water. The second term is the chemical accumulation due to consumption of sediment organic carbon. The third term represents the accumulation due to consumption of overlying phytoplankton. It can be noted that the BSF in general is a complicated function of sediment and overlying water dissolved and particulate chemical concentration, feeding rates and preferences, and the usual uptake, excretion, and growth rates.

Insight can be gained into the preceding equations by considering the following approximations. Let

$$\pi_s \approx K_{ow} \tag{38a}$$

$$N_{iw} \approx K_{ow} \tag{38b}$$

The first equation assumes that the sediment partition coefficient (on an organic carbon basis) is equivalent to the octanol-water partition coefficient, K_{ow}. The second of these equations assumes that growth effects and metabolism of contaminants are small and that the lipid-normalized BCF is also approximately equal to the octanol-water partitioning coefficient. From Equation 37 with $b_5 = 1.0$ (i.e., exposure to sediment only), the BSF is

$$S_5 \approx 1 + g_{5s} \tag{39}$$

Thus, if the sediment magnification factor, g_{5s} is small, the ratio of organism to sediment chemical concentration will be approximately unity, indicating no significant bioaccumulation from the sediment.

The BSF (S_3) for the forage fish (compartment #3) is given from Equation 10 as

$$S_3 = \frac{v_3}{r_s} = \frac{N_{3w}}{\pi_{ws}} + g_{32}S_2 + g_{35}S_5 \tag{40}$$

where S_2 is the BSF for the zooplankton. Note the inclusion of the sediment to water interaction given by π_{ws}.

The BAF for the forage fish (#3) with consumption of benthic organisms is given by (Thomann et al. 1992)

$$N_3 = \frac{v_3}{c_w} = \left\{ N_{3w} + g_{32}\left(N_{2w} + g_{21}N_{1w} \right) \right\}$$

$$+ \left\{ g_{35}\left[N_{5w}\left(b_{5s}\frac{\pi_{ws}}{\pi_s} + b_{5w} \right) + g_{5s}\pi_{ws} + g_{51}N_{1w} \right] \right\} \tag{41}$$

As seen, the first group of terms in braces is the BAF from a three-step food chain (see Equation 28b). The second term in braces is the accumulation due to consumption of benthic organisms.

Approximating Equation 41 by using Equation 38 and considering the forage fish to feed only on the benthic invertebrates ($g_{32} = 0$) and the invertebrates to be feeding on and exposed only to the sediment ($b_{5w} = 0$, $g_{51} = 0$) gives

$$N_3 = K_{ow} + \left(g_{35} + g_{35}g_{5s} \right)\pi_{ws} \tag{42}$$

The effect of the sediment interaction is now clearer. The BAF is elevated above the equilibrium level of K_{ow} by the magnitude of π_{ws} and the food chain effects. Equation 42 can be contrasted to a forage fish feeding exclusively in a pelagic chain. Thus

$$N_3 = K_{ow} + \left(g_{32} + g_{32}g_{21} \right)K_{ow} \tag{43}$$

Assuming the food chain multipliers are approximately similar, the impact of the sediment is the ratio of π_{ws} to K_{ow}.

Sediment-Water Column Interaction

The preceding equations for the BSF include the interaction between the water column and sediment. The ratio π_{ws} Equation 36 emerges when the benthic invertebrates are feeding on both sediment organic carbon and the carbon phytoplankton. The partitioning between the sediment and water column can be shown (Thomann et al. 1992) to be given at steady-state as

$$\pi_{ws} = \frac{\delta f_{ocw}\pi_w}{f_{ocs}} \tag{44}$$

where f_{ocs} and f_{ocw} are the sediment and water organic carbon fraction, respectively, and δ is a complicated function of settling and resuspension velocities of the particulates, net deposition velocity, sediment decay of chemical, and partitioning of chemical in water column and sediment (see Di Toro et al. 1982) given by

$$\delta = \frac{\left(v_u + v_d \right)f_{ps} + \left(\pi_s / \pi_w \right)K_f f_{ds} / \phi_s}{\left(v_u + v_d \right)f_{ps} + K_f f_{ds} / \phi_s + K_{ds}H_s} \tag{45}$$

where v_u and v_d are the resuspension and net deposition velocities (cm/yr), respectively, f_{ps} and f_{ds} are the fraction of chemical in particulate and dissolved

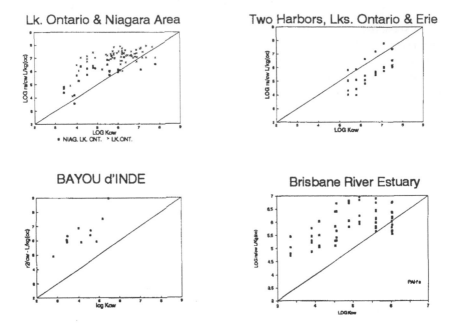

Figure 8. Variation of $r_s/c_{ws} = \pi_w s$ for several water bodies. Lake Ontario and Niagara area (Oliver and Niimi 1988, and Oliver and Charlton 1984), Two Harbors (Mudroch et al. 1989), Bayou d'Inde (Pereira et al. 1988), Brisbane River Estuary (Kayal and Connell 1990).

form, K_f is the interstitial diffusion rate (cm/d), ϕ_s is the sediment porosity and K_{ds} is the sediment decay rate (day^{-1}).

The water column partition coefficient is calculated using the formulation of Di Toro (1985) given by

$$\pi_w = \frac{K_{ow}}{1 + f_{ocw} K_{ow} m_w / 1.4} \tag{46}$$

for π_w in L/kg(oc) and m_w is the suspended solids concentration (kg/l).

Figure 8 shows the relationship of π_{ws} for several different areas and indicates that the partitioning is usually above K_{ow} levels. The preceding equations can be used to represent these data if sufficient information is available for the parameters such as net deposition velocity, suspended solids, etc.

CALIBRATION AND APPLICATION OF GENERIC MODELS
Pelagic Models

Applications of the generic model are given in Thomann (1981), (drawing on earlier work by Norstrom et al. 1976) and Servos (1988). In Connolly and

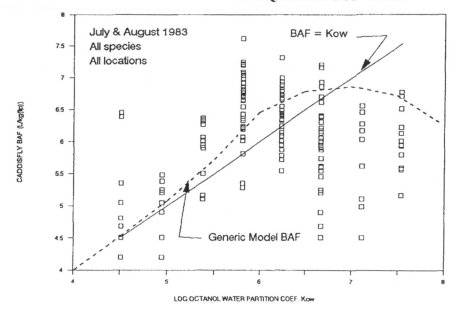

**Figure 9.   Comparison of generic model BAF to PCB congener data for
caddisfly in Upper Hudson River. Data compiled from Bush et
al. 1985 using percent lipid of 6.2%.**

Pedersen (1988) and Thomann (1989), a four-step food chain with parameters as
a function of K_{ow} was used. An additional application of Thomann (1989) is
shown in Figure 9 using the data of Bush (1985) on PCB congener accumulation
in caddisfly larvae in the upper Hudson River. Equation 28a was used, caddisfly
were assumed at 10 mg(w), 2% lipid, weight/dry weight = 5 and a food
conversion efficiency of 0.3. The phytoplankton BCF was assumed equal to K_{ow}
for log $K_{ow} \leq 6$. Above that level, the phytoplankton BCF was held constant at
10^6. Equation 18 was used to estimate chemical efficiency. The simple model as
shown in Figure 8 is a reasonable representation of the general trend of the data.
Bioaccumulation above BAF = K_{ow} is noted in the region of $5 \leq \log K_{ow} \leq 7$. The
decline above log K_{ow} of 7 is due to an assumed decreased chemical assimilation
efficiency.

Food Web Models with Sediment Interaction

The schematic for this model is shown in Figure 1. As noted earlier, an
important parameter in such a model is the sediment-water column interaction
given by π_{ws}. Data from Oliver and Charlton, (1984) for the Niagara River region
of Lake Ontario and Oliver and Niimi (1988) for the open Lake Ontario are used
to calibrate the relationship between sediment and overlying water concentrations. The data for Lake Ontario shown in Figure 8 were used for π_{ws}. (See
Thomann et al. 1992 for additional details).

The principal parameters that must be determined for the calibration of the food web-sediment model may divided into three groups. The first group of parameters are associated with the organism physiology including weight, food conversion efficiency, growth rate, and specific consumption rate. The second group of parameters reflect the behavior of the chemical and include the uptake and excretion rates and the chemical assimilation efficiency. The third group consists of the feeding preference factors. The parameter specifications for the first group follows that of Thomann (1989) where organism weight is assigned to each compartment, and allometric relationships are used to estimate growth and respiration and subsequent food consumption rates. Table 1 summarizes the parameters for the first group. Chemical uptake is calculated from weight and efficiency of transfer from the waterphase. The efficiency of aqueous uptake as well as the chemical assimilation efficiency of ingested prey is an assumed function of K_{ow} using Equation 18. Excretion is taken as the ratio of uptake to the octanol-water partition coefficient. Feeding preference parameters are determined from a sensitivity of the model using field data. Data from Oliver and Niimi (1988) are used for model calibration.

Figure 10 (top) from Thomann et al. (1992) shows the results of a sensitivity of the amphipod/sediment ratio given two assumptions on feeding behavior. For both A and B, the amphipod are assumed exposed to the overlying water only [i.e. $b_{5w} = 1.0$ in Equation 33]. Curve A indicates results when amphipods are assumed to feed exclusively on the sediment [$p_{5s} = 1.0$ in Equation 33]. Curve B assumes an exclusive phytoplankton diet [$p_{51} = 1.0$ in Equation 33]. As seen, the data appear to reflect some combination of feeding on sediment and overlying water particulates as well as exposure to interstitial and overlying water. Following analyses of combinations of feeding and water exposure, Figure 10 (bottom) shows the results of a 20% exposure of the benthic invertebrates to the interstitial water, 80% to overlying water chemical concentration and a 20% consumption of sediment particulate organic carbon. These combinations are not necessarily unique and other relationships between exposure and feeding may provide an equally credible representation of the observed data. The log K_{ow} region from about 6 to 8 shows a BSF greater than one indicating some bioaccumulation of chemical above an expected equilibrium value of about one (neglecting effects of organism growth). The values of v_5 of less than one are a result of exposure to overlying water chemical concentration.

Figure 11 from Thomann et al. (1992) shows the computed sculpin/sediment ratio compared to the observed data. The sculpin are considered a forage fish. For this computation, the amphipod parameters are from Figure 10 (bottom) and the sculpin are assumed to consume 80% benthic invertebrates and 20% zooplankton. The sediment/water column interaction is given from Figure 9. The particular shape of the model in Figure 11 is a result of a complicated interaction between contaminant exposure in the sediment and overlying water and chemical assimilation efficiency. The "water only" level is computed from Equation

Table 1.   Model Parameters for Food Web Pelagic/Sediment Model
(From Thomann et al. 1992)

Component No.	Weight (g(w))	f_{Li} % Lipid	Net Energy eff a^a	a_{wdi}	G Growth rate $(d^{-1})^b$	r Respiration rate $(d^{-1})^c$	$C_{L,i}$ Specific consumption[d]
1[f]	-	1	-	10	-	-	-
2	0.01	5	0.30	5	0.025	0.090	0.154
3	100	8	0.80	4	0.004	0.014	0.016
4	1000	20	0.80	4	0.0025	0.0090	0.0058
5	0.002	3	0.20	7	0.035	0.125	0.381
							1.52[e]

[a] g(OC) assim/g(OC) ingested (see Equation 22a).
[b] Equation 23.
[c] Equation 24.
[d] Equation 21 - g(1p)/g(1p)-d.
[e] Equation 22 - g(OC)/g(1p)-d.
[f] #1 = Phytoplankton, given by partitioning to water concentration.

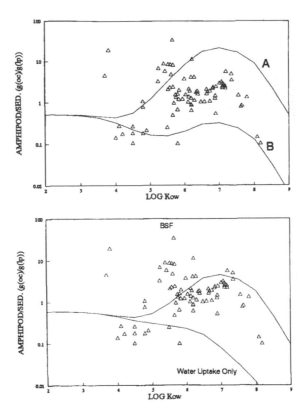

Figure 10.   Comparison of observed data of Oliver and Niimi (1988) to benthic invertebrate accumulation model using E(33). See text for discussion.

40 with zero feeding. As seen, above a log K_{ow} of about five, this model indicates that virtually all of the sculpin chemical concentration is due to uptake from the food route.

SITE-SPECIFIC MODEL

Site-specific models incorporate all of the preceding mechanisms with various levels of the food chain as functions of time and age (see e.g., Thomann and Connolly 1984, Connolly and Tonelli 1985, and Connolly 1990). Two additional applications are discussed here.

PCB-Hudson Estuary Striped Bass

The food chain accumulation of PCBs in the striped bass of the Hudson Estuary has been evaluated by Thomann et al. (1989, 1991). An age-dependent

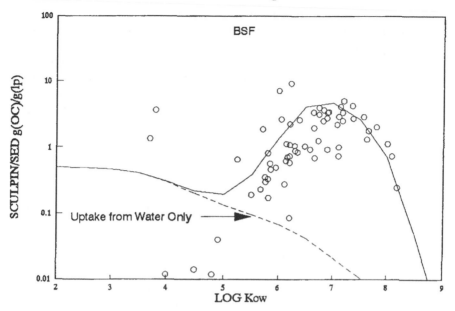

**Figure 11.   Comparison of observed data of Oliver and Niimi (1988) to
sculpin chemical data using Equation 40. Solid line ≈ total
accumulation from food and water.**

model (see Figure 2) was used and calibrated to the striped bass residue data. An
important part of that model is the determination of the food web interactions
used to construct model linkages. These interactions are reviewed here as an
example of the factors that must be considered in a site-specific model.

Food Web Interactions

The determination of the food web interactions for the striped bass begins
with an evaluation of feeding patterns by various age classes. A variety of feeding
studies have been conducted where stomach contents of the striped bass are
examined, typed, and enumerated. O'Connor (1984), Gardinier and Hoff (1982),
and Setzler et al. (1980) provide a basis for estimating the diet of the striped bass.
An earlier general survey of striped bass life history is given by Raney (1952).

In general, these studies indicate that the striped bass progress from primarily
invertebrate feeders in age classes zero to one to two years to primarily
piscivorous feeders in older age classes.

For striped bass in age class zero to one+, collected during 1974 to 1977,
Gardinier and Hoff (1982) indicate a diet of primarily invertebrates. Major
groups included *Gammarus*, calanoid copepods, and cladocerans. Gardinier and
Hoff (1980) indicate that 80 to 100% of the diet of "small striped bass" was
Gammarus fasciatus. O'Connor (1984) in a survey of striped bass stomach

contents in the New York Harbor region in the winter of 1983 also concluded that for fish, the principal component of the diet was *Gammarus* although other taxa were also present.

For age class zero+ to about two years, Gardinier and Hoff (1980) reported a mixed diet of fish (blueback herring, tomcod, bay anchovy and mummichog) and invertebrates. Seasonal variations were significant. In June, fish made up 67% of the total diet, but by September, invertebrates comprised 83% of the total diet. For Long Island Sound, Setzler et al. (1980) summarizing the work of Schaefer (1970) indicated that 85% of the food volume of striped bass of age less than about three years consisted of invertebrates. *Gammarus* and mysis shrimp *Neomysis americanca* were dominant.

For age classes greater than about two+, the striped bass are almost entirely piscivorous. The percentage of fish remains high with increasing size of the striped bass. For striped bass between 200 to 399 mm, fish made up 31% of the stomach contents. For striped bass greater than 800 mm (> about eight years old), fish made up about 86% of the stomach contents. A similar result is summarized by Seltzer et al. (1980) who indicate that fish make up greater than 65% of the diet of striped bass greater than six years old. The fish consumed include white perch, Atlantic tomcod, blueback herring, and spottail shiner.

The lower levels of the food web are considered to be composed of a "phytoplankton" compartment which is considered to be in a sorption-desorption equilibrium state with the dissolved water concentration of a given PCB homolog. The phytoplankton are then preyed upon by a "zooplankton" compartment, the characteristics of which are considered to be represented by *Gammarus*.

The next level is assigned as a representation of "small fish", a compartment meant to reflect a mixed diet of fish of about 10 g in weight. This compartment would therefore include such fish as zero to one+ tomcod and herring. The white perch are considered to feed exclusively on zooplankton.

The striped bass is divided into seventeen age classes, the characteristics of which are discussed more fully below. For classes zero to two years, the striped bass are assumed to feed on the zooplankton (i.e., *Gammarus*). For age classes from two to six years, the striped bass are assumed to feed on a mixture of small fish and age class yearling white perch. For striped bass older than six years, white perch from two to five years old are assumed to comprise the diet.

The total number of state variables then in the food web model is 29, since each age class of a species represents a separate state variable.

The calibrated striped bass model indicated significant bioaccumulation as shown in Figure 11. The top figure shows the calculated time history of total PCB ($\mu g/g/(lp)$) in the striped bass for the mid-Hudson region over a 40-year period. (A physico-chemical model was used to calculate the time variable water column concentrations; see Thomann et al. 1991). As shown, better than 90% of the total PCB residue is calculated to be due to the food chain transfers depicted in Figure 2. Figure 12 (bottom) shows the calculated contribution due to water only

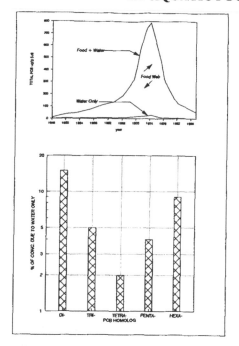

Figure 12.   (Top) Contribution to total PCB concentration in Hudson
River striped bass age dependent, time dependent model from
food web and water only. (Bottom) Variation of uptake due to
water only as a function of PCB homolog group. From Thomann
et al. (1989).

exposure as a function of PCB homolog. The variation is due largely to the
assumed variation in chemical assimilation efficiency which was homolog
dependent. For the tetra PCBs, only 2% of the striped bass chemical concentration is considered to be due to aqueous exposure. This percentage rises to about
10% for the less assimilable lower and higher chlorinated PCBs.

An important step in determining the ecosystem response to various load
simulations is to analyze the behavior of the mean concentration of a top predator
as a function of the percent below a certain target level of, say, $2\,\mu g/g(w)$ for PCBs.
The question addressed for the Hudson striped bass was:

What is the estimated mean concentration in the striped bass (over the sampled
age class of three to six years) so that a given percentile of the fish will be below
$2\,\mu g/g(w)$?

A log normal distribution of the striped bass PCB concentration is assumed
based on the analysis of the existing data (Thomann et al. 1989). The only other

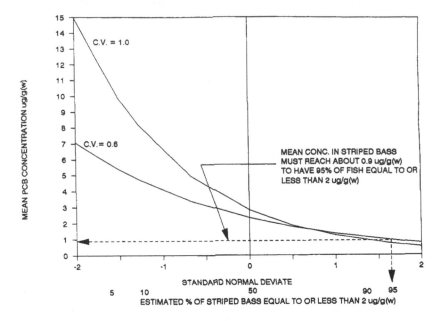

Figure 13. Relationship between mean total PCB concentration in Hudson River striped bass and percentiles equal to or less than 2 μg/g(w) from Equations 47–49.

parameter to be specified is the coefficient of variation of the log normally distributed variable. The equations are as follows:

$$\sigma_{\ln x}^2 = \ln\left[1 + \delta_x^2\right] \tag{47}$$

$$\mu_{\ln x} = \ln x - \sigma_{\ln x} \cdot z \tag{48}$$

$$\mu_x = \exp\left[\mu_{\ln x} + \frac{1}{2}\sigma_{\ln x}^2\right] \tag{49}$$

where $\sigma_{\ln x}^2$ is the variance of log transformed fish concentration, δ_x is the coefficient of variation, $\mu_{\ln x}$ is the log transformed mean concentration, ln x is the logarithm of the target concentration (here set equal to 2 μg/g(w)), Z is the standardized normal deviate, and μ_x is the required mean fish PCB concentration.

Figure 13 shows the relationship between the percentile of striped bass concentration and the mean concentration for coefficients of variation of 0.6 and 1.0. These values are approximately the range of coefficients as obtained from the measured data.

As indicated in Figure 13, if a mean concentration of about 0.9 μg/g(w) is reached in the striped bass, 95% of the fish would be expected to have concentrations below 2 μg/g(w). A mean concentration of about 3 μg/g(w) would result in about 50% of the fish below 2 μg/g(w). It can also be noted that the difference between the two coefficients of variation is not significant for percentiles greater than 50%. The results of this analysis as displayed in the figure were used to estimate the time it will take to reach various percentiles of PCB concentration.

LAKE ONTARIO PCB
Modeling the PCB Time History in the Lake Ontario Lake Trout Food Chain

As part of an International Joint Commission (IJC) study to compare various modeling frameworks a model of the fate of PCBs in the water column, sediment, and biota of Lake Ontario was developed (Connolly et al. 1987). The lake was treated as a completely mixed water column overlying a vertically segmented sediment. The model simulated dissolved and particulate PCB concentrations over the period 1943 to 1981. Measured water column total PCB concentrations in 1980 and 1981 and measured surficial sediment PCB concentrations in 1968, 1972, 1979, and 1981 were used in model calibration. The computed water column dissolved PCB and sediment dissolved and particulate PCB defined exposure time histories that were used to compute the PCB time history in the lake trout food chain.

The food chain model used was essentially that previously developed to model PCB in the lake trout food chain of Lake Michigan (Thomann and Connolly 1984, Thomann et al. 1987). The species included, in addition to lake trout, were the alewife, the pelagic invertebrate *Mysis relicta*, the benthic invertebrate *Pontoporeia hoyi* and phytoplankton. *Pontoporeia* are deposit feeders and thus provide a vector for transferring sediment PCB to the upper levels of the food chain.

The Lake Michigan model was modified to make the PCB excretion rate a function of the animal's lipid content. Consistent with laboratory data the excretion rate was assumed to be inversely related to the fraction lipid of the animal (f_L). Specifically, the excretion rate constant was computed as follows:

$$K = \frac{k_u}{f_L K_{ow}} \qquad (50)$$

where k_u is the uptake rate constant at the gill and K_{ow} is the octanol-water partition coefficient for PCB (log K_{ow} was assumed to be 6.3).

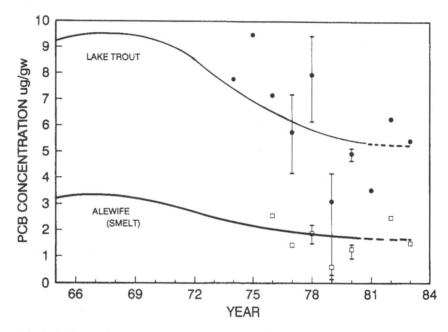

**Figure 14.   Calculated and observed time history of total PCB in the Lake
Ontario lake trout food chain (Connolly et al. 1987).**

Using a PCB partition coefficient of 30 L/g(w) for phytoplankton and an
alewife PCB assimilation efficiency of 0.6 (rather than the 0.7 value used in the
Lake Michigan calibration) the model reproduced the time history of both lake
trout and alewife (data for smelt included) PCB concentrations fairly well
(Figure 14). As seen in the figure, the computed average lake trout concentration
reaches a peak of about 9.5 µg/g(w) in the late 1960s and declines to about 5.5
µg/g(w) in 1980. The maximum PCB concentration occurs in 11-year old trout.
Their computed average concentration in 1980 is 9.7 µg/g(w). A similar pattern
is seen for alewife. By 1980 their average concentration has declined to about 2
µg/g(w), with an average maximum concentration of 3.3 µg/g(w) occurring in
six-year old alewife.

ACKNOWLEDGEMENTS

Grateful appreciation is offered to our colleague Dominic M. Di Toro for his
always insightful comments. Thanks are also given to Eileen Lutomski for her
patient typing of the manuscript.
    This work was supported under a research grant from NIEHS, Grant #
1P42ES04895 and by a cooperative agreement between the U.S. EPA and
Manhattan College.

REFERENCES

Barber, M.C., L.A. Suarez and R.R. Lassiter. "Modeling Bioconcentration of Nonpolar Organic Pollutants by Fish," *Environ. Toxicol. Chem.* 7:545-558 (1988).

Bierman, V. "Equilibrium Partitioning and Biomagnification of Organic Chemicals in Benthic Animals," *Environ. Sci. Technol.* 24(9):1407-1412(1990).

Bush, B., K.W. Simpson, L. Shane and R.R. Koblintz. "PCB Congener Analysis of Water and Caddisfly Larvae (Insecta: Trichoptera) in the Upper Hudson River by Glass Capillary Chromatography," *Bull. Environ. Contam. Toxicol.* 34:96-105(1985).

Connolly, J.P. "Application of a Food Chain Model to PCB Contamination of the Lobster and Winter Flounder Food Chains in New Bedford Harbor," *Environ. Sci. Technol.* 25(4):760-769 (1991).

Connolly, J.P., D.M. Di Toro, C.J. Pedersen and J.R. Newton. "A Model of PCB in the Water, Bed and Food Chain of Lake Ontario," U.S. EPA report, Region V, Chicago, IL, part of the International Joint Commission Task Force on Chemical Loading (1987).

Connolly, J.P., and C. Pedersen. "A Thermodynamic-Based Evaluation of Organic Chemical Accumulation in Aquatic Organisms," *Environ. Sci. Technol.* 22(1):99-103(1988).

Connolly, J.P., and R. Tonelli. "Modelling Kepone in the Striped Bass Food Chain of the James River Estury," *Estuarine, Coastal Shelf Sci.*, 20:349-366(1985).

Connor, M.S. "Fish/Sediment Concentration Ratios for Organic Compounds," *Environ. Sci. Technol.* 18:31-35(1984).

Di Toro, D.M. "A Particle Interaction Model of Reversible Organic Chemical Sorption." *Chemosphere* 14(10):1503-1538(1985).

Erickson, R.J., and J.M. McKim. "A Simple Flow-Limited Model for Exchange of Organic Chemicals at Fish Gills," *Environ. Toxicol. Chem.* 9:159-165(1990).

Fowler, S.W., G.G. Polikarpov, D.L. Elder, P. Parsi and J.P. Villeneuve. "Polychlorinated Biphenyls: Accumulation from Contaminated Sediments and Water by the Polychaete *Nereis diversicolor*," *Mar. Biol.* 48:303-309(1978).

Gardinier, M.N., and T.B. Hoff. "Diet of Striped Bass in the Hudson River Estuary," *NY Fish Game Jour.* 29(2):152-165(1982).

Gobas, F.A.P.C., A. Opperhuizen and O. Hutzinger. "Bioconcentration of Hydrophobic Chemicals in Fish: Relationship with Membrane Permeation," *Environ. Toxicol. Chem.* 5:637-646(1986).

Gobas, F.A.P.C., and D. Mackay. "Dynamics of Hydrophobic Organic Chemical Bioconcentration in Fish," *Environ. Toxicol. Chem.* 7:545-558(1987).

Gruger, E.H., N.L. Karrick, A.I. Davidson and T. Hruby. "Accumulation of 3,4,3',4'-Tetrachlorobiphenyl and 2,4,5,2',4',5'-and 2,4,6,2',4',6'-Hexachlorobiphenyl in Juvenile Coho Salmon," *Environ. Sci. Technol.* 9(2)121-127(1975).

Hallam, T.G., R.R. Lassiter and S.A.L.M. Kooijman. *Effects of Toxicants on Aquatic Populations in Applied Mathematical Ecology.* S.A. Levin, T.G. Hallam and L.J. Gross, Eds., (New York: Springer-Verlag, 1989) pp. 352-382.

Huckins, J.N., T.R. Schwartz, J.D. Petty and L.M. Smith. "Determination, Fate, and Potential of PCBs in Fish and Sediment Samples with Emphasis on Selected AHH-Inducing Congeners." *Chemosphere* 17:1995-2016(1988).

Karickhoff, S.W. "Organic Pollution Sorption in Aquatic Systems," *Am. Soc. Civ. Eng. J. Hydraul. Div.* 10(6):707-735(1984).

Karickhoff, S.W., D.S. Brown and T.A. Scott. "Sorption of Hydrophobic Pollutants on Natural Sediments," *Water Res.* 13:241-248(1979).

Kayal, S.I., and D.W. Connell. "Partitioning of Unsubstituted Polycyclic Aromatic Hydrocarbons Between Surface Sediments and the Water Column in the Brisbane River Estuary," *Aust. J. Mar. Freshwater Res.* 41:443-456(1990).

Landrum, P.F., W.R. Faust and B.J. Eadie. "Bioavailability and Toxicity of a Mixture of Sediment-Associated Chlorinated Hydrocarbons to the Amphipod *Pontoporeia hoyi.*," in *Aquatic Toxicology and Hazard Assessment*; Vol. 12, ASTM STP 1027, U.M. Cogwill and L.R. Williams, Eds., (Phila, PA: Amer. Soc. Testing & Materials, 1989).

Lederman, T.C., and G.-Y. Rhee. "Bioconcentration of a Hexchlorbiphenyl in Great Lakes Planktonic Algae," *Can. J. Fish. Aquat. Sci.* 39(3): 380-387(1982).

Mackay, D.. "Correlation of Bioconcentration Factors," *Environ. Sci. Technol.* 16:274-278(1982).

Malins, D.C., B.B. McCain, D.W. Brown, S. Chan, M.S. Myers, J.T. Landahl, P.G. Prohaska, A.J. Friedman, L.D. Rhodes, D.G. Burrows, W.D. Gronlund and H.O. Hodgins. "Chemical Pollutants in Sediments and Diseases of Bottom-Dwelling Fish in Puget Sound, Washington," *Environ. Sci. Technol.* 18:705-713(1984).

McKim, J., P. Schmieder and G. Veith. "Absorption Dynamics of Organic Chemical Transport Across Trout Gills as Related to Octanol-Water Partition Coefficient," *Toxicol. Appl. Pharmacol.* 77:1-10(1985).

Mudroch, A., F.I. Onuska and L. Kalas. "Distribution of Polychlorinated Biphenyls in Water, Sediment and Biota of Two Harbours," *Chemosphere* 18:2141-2154(1989).

Nalepa, T.F., and P.F. Landrum. "Benthic Invertebrates and Contaminant Levels in the Great Lakes: Effects, Fates, and Role in Cycling," in *Toxic Contaminants and Ecosystem Health: A Great Lakes Focus*, M.S. Evans, Ed. (New York: J. Wiley & Sons, 1988).

Norstrom, R.J., A.E. McKinnon and A.S.W. DeFreitas. "A Bioenergetics-Based Model for Pollutant Accumulation by Fish. Simulation of PCB and Methylmercury Residue Levels in Ottawa River Yellow Perch (*Perca flavescens*)," *J. Fish. Res. Board Can.* 33:248-267(1976).

O'Connor, J.M. "PCBs: Dietary Dose and Burdens in Striped Bass from the Hudson River," *Northeast. Environ. Sci.* 3(3/4):152-158(1984).

Oliver, B.G. "Biouptake of Chlorinated Hydrocarbons from Laboratory-Spiked and Field Sediments by Oligochaete Worms," *Environ. Sci. Technol.* 21:785-790(1987).

Oliver, B.G., and A.J. Niimi. "Trophodynamic Analysis of Polychlorinated Biphenyl Congeners and Other Chlorinated Hydrocarbons in the Lake Ontario Ecosystem," *Environ. Sci. Technol.* 22:388-397(1988).

Oliver, B.G., and M.N. Charlton. "Chlorinated Organic Contaminants on Settling Particulates in the Niagra River Vicinity of Lake Ontario," *Environ. Sci. Technol.* 18:903-908(1984).

Parkerton, T.F. "Modeling the Bioaccumulation of Chlorinated Hydrocarbons in a Detroit River Benthic Foodchain," Manuscript in preparation (1991).

Pereira, W.E., C.E. Rostad, C.T. Chiou, T.I. Brinton and L.B. Barber, II. "Contamination of Estuarine Water, Biota, and Sediment by Halogenated Organic Compounds; A Field Study," *Environ. Sci. Technol.* 22:772-778(1988).

Poje, G.V., S.A. Riordan and J.M. O'Connor. "Food Habits of the Amphipod *Gammarus tigrinus* in the Hudson River and the Effects of Diet upon its Growth and Reproduction," in *Fisheries Research in the Hudson River*, C.L. Smith, Ed., (Albany, NY: State University of New York Press, 1988), pp. 255-270.

Raney, E.C. "The Life History of the Striped Bass, *Roccus saxatilis* (Waldbaum)," *Bull. of Bingham Ocean, Collection.* XIV:1-97(1952).

Rubenstein, N.I., W.T. Gilliam and N.R. Gregory. "Dietary Accumulation of PCBs from a Contaminated Sediment Source by a Demersal Fish (*Leiostomus xanthurus*)," *Aquat. Toxicol.* 5:331-342(1984).

Servos, M.R. "Fate and Bioavailability of Polychlorinated Dibenzo-p-dioxins in Aquatic Environments," Ph.D. thesis, University of Manitoba, Winnipeg, Manitoba, Canada (1988).

Setzler, E.M., W.R. Boynton, K.V. Wood, H.H. Zion, L. Lubbers, N.K. Mountford, P. Fere, L. Tucker and J.A. Milhursky. "Synopsis of Biological Data on Striped Bass, *Morone saxatilis* (Waldbaum)," NOAA Technical Report, NMFS Cir. 433, FAO Synopsis No. 121, U.S. Dept. of Commerce, Rockville, MD (1980), p. 69.

Thomann, R.V. "Equilibrium Model of Fate of Microcontaminants in Diverse Aquatic Food Chains," *Can. J. Fish. Aquat. Sci.* 38:280-296(1981).

Thomann, R.V., and J.P. Connolly. "Model of PCB in the Lake Michigan Lake Trout Food Chain," *Environ. Sci. Technol.* 18:65-71(1984).

Thomann, R.V. "Bioaccumulation Model of Organic Chemical Distribution in Aquatic Food Chains," *Environ. Sci. Technol.* 23(6):699-707(1989).

Thomann, R.V., J.A. Mueller, R.P. Winfield and C.-R. Huang. "Mathematical Model of the Long-Term Behavior of PCBs in the Hudson River Estuary," Final Report to Hudson River Foundation, Ch. 8, Appendix (1989).

Thomann, R.V., J.P. Connolly and N.A. Thomas. "The Great Lakes Ecosystem — Modeling the Fate of PCBs," in *PCBs and the Environment, Vol. 3*, J.S. Waid, Ed., (Boca Raton, FL: CRC Press, Inc., 1987), pp. 153-180.

Thomann, R.V., J.A. Mueller, R.P. Winfield and C.-R. Huang. "Model of PCB Homologs in the Hudson Estuary," *ASCE, Environ. Eng. Div.* 117(2):161-178 (1991).

Thomann, R.V., J.P. Connolly and T.F. Parkerton "An Equilibrium Model of Organic Chemical Accumulation in Aquatic Foodwebs with Sediment Interaction," Manuscript accepted for publication, *Environ. Toxicol. Chem.* 11:615-629 (1992).

U.S. EPA. "Briefing Report to the EPA Science Advisory Board on the Equilibrium Partitioning Approach to Generating Sediment Quality Criteria," EPA Report-440/5-89-002, Office of Water Regulations and Standards, Criteria and Standards, Washington, D.C., 8 Sects. (1989).

Wilkinson, L. *SYSTAT: The System for Statistics.* (Evanston, IL: SYSTAT, Inc., 1988), p. 822.

8 DERIVATION OF BIOACCUMULATION PARAMETERS AND APPLICATION OF FOOD CHAIN MODELS FOR CHLORINATED DIOXINS AND FURANS

INTRODUCTION

The presence of 2,3,7,8-tetrachlorodibenzo-*p*-dioxin (TCDD) and 2,3,7,8-tetrachlorodibenzofuran (TCDF), as well as other chlorinated dioxins (PCDDs) and furans (PCDFs) in aquatic biota from freshwater and marine habitats near bleached kraft mills has recently been well-documented (Whittle et al. 1990, Harding and Pomeroy 1990, Rogers et al. 1989, Mah et al. 1989, Kuehl et al. 1987,1989). The pathways by which PCDDs and PCDFs are transferred from mill effluent to fish are not well-understood. TCDD and TCDF are associated with suspended and dissolved organic carbon in mill effluents (Amendola et al. 1989, Clement et al. 1989) and river water (Merriman 1988). Association of PCDDs with dissolved and particulate organic carbon has been shown to reduce the bioavailability to fish via direct uptake from water (Servos et al. 1989). Steric factors and biotransformation (of non-2,3,7,8-substituted congeners) also influence the apparent uptake of PCDD/Fs from water (Opperhuizen and Sijm 1990). But PCDDs and PCDFs are assimilated from ingested sediment particles by fish (Kuehl et al. 1987, Cook et al. 1990, Van der Weiden et al. 1989) and invertebrates (Rubenstein et al. 1990). Assimilation efficiencies from fish food of about 50% for 2,3,7,8-TCDD (Kleeman et al. 1986) and 42% for 2,3,4,7,8-PnCDF have been found in juvenile rainbow trout (Muir et al. 1990). Highly chlorinated PCDD/Fs are poorly accumulated from water but are accumulated via the gut food possibly because steric factors are not as critical to permeation

ISBN 0-87371-511-X

of the gut membrane as they are with gill membranes (Opperhuizen and Sijm 1990).

Food chain transfer has been recognized as the most important pathway of accumulation of hydrophobic organochlorine pollutants such as PCBs (Thomann 1981, 1989, Thomann and Connolly 1984, Clark et al. 1990) and 2,3,7,8-TCDD (Cook et al. 1990, Endicott et al. 1990). Application of food chain models requires knowledge of food web relationships, growth rates, feeding rates, and pharmacokinetic parameters combined with estimates of pollutant concentrations in water and suspended particles (Thomann et al. 1991, this volume, Thomann and Connolly 1984, Connolly and Tonelli 1985). For very hydrophobic compounds assimilation efficiency at each trophic level, and sorption by detritus/phytoplankton, are critical to the prediction of food chain transfer (Thomann 1989).

This paper summarizes the results of our studies in which pharmacokinetic parameters have been derived for use in food chain modeling of PCDD/F's concentrations in fish and invertebrates. Laboratory dietary exposure studies were conducted with fish to determine assimilation and elimination of PCDFs and PCDDs. For invertebrates we have used natural populations in lake mesocosms treated with ^{3}H-2,3,7,8-TCDF, located littoral zone of a small Canadian shield lake. The ultimate goal of these studies was to predict the accumulation of 2,3,7,8-TCDD and TCDF in fish from rivers and lakes near bleached kraft pulp mill discharges. Two modeling approaches for predicting food chain transfer at steady-state were compared. The first is based on the work of Thomann (1981) and calculates concentrations in fish at steady-state based on experimentally determined pharmcokinetic parameters. The second food chain model, which is discussed in more detail in this volume (Gobas 1991) makes similar calculations but requires only information regarding size, lipid levels, and feeding relationships.

METHODS
Pharmacokinetic Studies

Assimilation and depuration of PCDD/F's were studied using juvenile rainbow trout (2 to 4 g initial weight) held at 10°C in a continuous flow of dechlorinated (UV-carbon filtered) tap water. Exposure conditions such as lighting period, water flow, and tank size were virtually identical in all studies. The trout were fed commercial fish food (Martin's Feed Mills Ltd., Elmira Ont. (41% protein, 15% lipid, 3% fiber) daily, containing 0.36 to 50 ng g^{-1} of each PCDF/PCDD congener (^{14}C-or ^{3}H-labelled and nonlabelled) at 0.015 to 0.02 g food·g fish^{-1}·d^{-1} for 30 days. The compounds studied are listed in Table 1. At each sampling time the feeding rate was adjusted for fish growth. Fish were sampled at five-day intervals during the uptake phase of the experiments. Depuration in remaining fish was studied for up to 180 days while feeding food without added PCDFs.

Table 1.   Depuration Half-Lives and Equilibrium Factors (Lipid Basis) for PCDD/Fs in Rainbow Trout Following 30 Day Dietary Exposure[a]

Congener	Food conc'n ng g^{-1}	Depuration[b] half-life days	Assimilation[c] efficiency %	Biomagnification factor[d]
		PCDDs[d]		
1,2,3,7-TCDD	110	4 ± 1	14 ± 1.8	0.06
1,3,6,8-TCDD	113	5 ± 1	13 ± 3.7	0.03
1,2,3,4,7-PnCDD	105	2 ± 2	19 ± 2.3	0.08
1,2,3,4,7,8-HxCDD	109	44 ± 6	37 ± 2.7	0.50
1,2,3,4,6,7,8-HpCDD	109	39 ± 9	13 ± 1.3	0.27
OCDD	94	13 ± 12	18 ± 17	0.05
		PCDFs[d]		
2,8-DCDF	100	10 ± 2	25 ± 12	0.05
1,2,3,7-TCDF	50	6 ± 3	23 ± 20	0.03
1,2,7,8-TCDF	50	3 ± 1	10 ± 4	0.01
2,3,7,8-TCDF	0.36	73 ± 5	59 ± 3	2.02
	7.2	70 ± 3	55 ± 3	1.80
1,2,3,4,8-PnCDF	50	20 ± 5	10 ± 6	0.03
2,3,4,7,8-PnCDF	0.82	69 ± 5	41 ± 2.2	1.44
	9.0	61 ± 5	44 ± 1.6	1.33
1,2,4,6,7,8-HxCDF	50	20 ± 5	27 ± 7	0.12
1,2,4,6,7,8-HxCDF	50	14 ± 6	25 ± 12	0.08
OCDF	100	11 ± 7	18 ± 11	0.04

[a] Calculated from the log linear portion of a first order decay curve. $T_{1/2}=0.693/k_2$ ($\pm$ standard error) where k_2 is the slope of the depuration curve.

[b] Assimilation efficiency (E) calculated from the uptake curve (growth corrected concentrations on lipid weight basis in whole fish versus time in days) using iterative nonlinear regression.

[c] Equilibrium biomagnification factor: $BMF = E \cdot F/k_2$, where F = feeding rate (g lipid $\cdot$g fish lipid$^{-1} \cdot$d^{-1}).

[d] Results for PCDDs from Muir and Yarechewski (1988) and for 2,3,4,7,8-PnCDF from Muir et al. (1990).

Whole fish were freeze-dried and extracted by blending with toluene using a Polytron (Brinkmann Instruments) homogenizer. PCDF and PCDD were quantified by counting the extract directly using liquid scintillation counting (LSC) and, after separation from co-extractive lipids, by gas chromatography/mass spectrometry (Hewlett-Packard 5980 GC-5970 MSD) using selected ion monitoring. The lipid content of each fish was determined gravimetrically after drying a portion of the extract to constant weight. Further details on sampling and analysis are given by Muir and Yarechewski (1988) and Muir et al. (1990).

Pharmacokinetic parameters required for the food chain model of Thomann (1981) were obtained from the lipid-normalized accumulation data after correction for growth dilution. Elimination rate constants (k_2) were obtained by fitting data from the depuration phase to a first-order decay curve. The best value of assimilation (E) was calculated by fitting the data to the integrated form of the kinetic rate equation for constant dietary exposure (Bruggeman et al. 1981) using iterative nonlinear regression:

$$C_{fish} = \left(E \cdot F \cdot C_{food} / k_2 \right) \cdot \left(1 - \exp\left(-k_2 t \right) \right) \tag{1}$$

where

C_{fish} = concentration in fish (lipid weight) at time t (days)
C_{food} = concentration in food (lipid weight)
F = feeding rate (g food·g fish^{-1}·d^{-1}) on a lipid weight basis

Mesocosm Studies with 2,3,7,8-TCDF

Limnocorrals (5 m diameter, 2 meter depth, 34 m^3 volume) were installed in the littoral zone of Lake 375 a small (21 ha) oligotrophic Canadian shield lake in Experimental Lakes Area in Northwestern Ontario. They consisted of a woven plastic tube with styrofoam flotation collar anchored firmly to the sediment. Bottom sediments (0 to 2 cm layer) were 80 to 90% sand, < 10% coarse (> 2 mm) granitic material, from 5 to 10% silt and < 5% clay by weight. Surface layers had a thin (about 1 cm) covering of organic floc. Sediments were about 1% total organic carbon and 0.1% total nitrogen on a dry weight basis. A natural benthic invertebrate community was enclosed within the mesocosm. ^{3}H-TCDF was added to four of six enclosures sorbed to 10 g sediment per limnocorral. Further details on the addition of TCDF are given by Fairchild et al. (1991).

Water, suspended particles (0.2 to 20 µm), sediment cores, pore water, benthic invertebrates, and emerging inserts were sampled weekly for the first three weeks after addition of TCDF, then after 60 and 120 days. Critical to this discussion are the results for sediment, benthic invertebrates, and emerging insects. Complete results are reported elsewhere (Fairchild et al. 1991). Sediment core samples were extracted and the extracts analysed by LSC as described for fish. Single or pooled samples of Chironomid larvae, *Hexagenia* nymphs, and emerging insects were combusted on a Packard 306 oxidizer then assayed for ^{3}H$_2$O by LSC. A lipid content of 23% (dry weight) for larvae and emerging insects, and 17% (dry weight for *Hexagenia*, was assumed based on literature values (Herbes and Allen 1983). Water samples (4l) were extracted with dichloromethane and the extract was counted by LSC. Concentrations of dissolved and organic carbon-associated TCDF concentrations in water and sediment pore water were determined by centrifuging at 12,000 g, in stainless steel bottles, and extracting the supernatant.

Values of E and k_2 for chironomids, *Hexagenia* and emerging insects were derived by fitting the TCDF accumulation data to Equation 1 using iterative nonlinear regression with two unknown parameters (E and k_2)

Food Chain Models

A steady-state version of the "generic" equilibrium food chain model of Thomann (1981) and Thomann et al. (1991, this volume), consisting of three compartments, was used to simulate the food chain in rivers near pulp mill. The trophic levels were organic detritus/biosolids (< 20 μm) as the base of the food chain, particle feeding invertebrates (e.g., chironomid larvae, *Hexagenia* nymphs), and adult fish feeding on benthic larval invertebrates and emerging adults. At steady-state the concentration at each tropic level i, $C_{F,i}$ is given by

$$C_{F,i} = \left(k_{1,i}C_w + E \cdot F \cdot C_{P,i-1}\right) / k_2 \qquad (2)$$

where

C_w = concentration in the water (ng·l^{-1})
$C_{F,i}$ = concentration at each trophic level (ng kg^{-1})
$C_{p,i-1}$ = concentration in the prey species (ng kg^{-1})
$k_{1,i}$ = the rate constant for uptake from water (units l·kg^{-1}·d^{-1})
k_2 = elimination rate constant (d^{-1})

In the absence of field data for TCDD/F concentrations in the dissolved phase in receiving waters (C_w), levels in water were obtained by dividing reported concentrations of 2,3,7,8-TCDD and TCDF in mill effluents (PPRIC 1989,1991) by the average dilution factor for each mill effluent (Haliburton 1990). Concentrations in organic detritus and biosolids were derived from the Karickhoff (1979) relationship, $K_{oc} = 0.411·K_{ow}$, where K_{oc} is the organic carbon sorption coefficient and K_{ow} is the octanol-water partition coefficient.

A key assumption in this model is that the pathway of TCDD/F from mill effluent to fish in riverine ecosystems is via dilution and sorption to particulates. Bed sediments and pore waters were not included in the model. Benthic invertebrates were assumed to accumulate TCDD/F from particulate detritus/ biosolids and from the dissolved phase of the overlying water. Benthic feeding fish were assumed to accumulate TCDD/F from invertebrates and from the dissolved phase. Mountain whitefish (*Prosopium williamsoni*) and lake white-fish (*Coregonus clupeoformus*) were assumed to consume only benthic inverte-brates while suckers (*Catostomus* sp) were assumed to accumulate TCDD and TCDF from detritus biosolids (25%) and invertebrates (75%).

The food chain accumulation model of Gobas (1991, this volume) was used with the same three-level food chain described above. The model derives internally the rate constants of chemical uptake from water, uptake from food consumption, elimination via the gills to water, elimination by faecal egestion, and the rate constant of growth dilution, based on fish weight, lipid content and K_{ow}. Trophic transfer and magnification were then calculated at steady-state from feeding preferences which are also entered into the model.

At steady-state the concentrations in fish are given by

$$C_f = \left(k_1 \cdot C_w + k_D \cdot \left(\Sigma_i P_i \cdot C_{D,i}\right)\right) / \left(k_2 + k_E + k_G + k_M\right) \qquad (3)$$

where

- k_D = rate constant for chemical uptake from food (kg food·kg fish^{-1}·d$^{-1)}$
- k_E = rate constant for chemical elimination by faecal egestion (d^{-1})
- k_M = rate constant for metabolic transformation (d^{-1})
- k_G = rate constant for growth dilution (d^{-1})
- P_i = fraction of the diet of the fish consisting of prey i
- $C_{D,i}$ = chemical concentration in prey i (μg/kg^{-1})
- C_W = bioavailable or dissolved water concentration (ng/l^{-1})

The concentration in the benthos C_B (μg kg^{-1}) is calculated as

$$C_B = L_B \cdot d_L \cdot C_S / X_S \cdot d_{OC} \qquad (4)$$

where

- L_B = lipid fraction of the benthos
- d_L = density of lipids (kg/l^{-1})
- X_S = organic carbon fraction of sediments
- d_{OC} = density of the sediment organic matter
- C_s = concentration in sediment

The model assumes that benthos is in chemical equilibrium with sediments and pore water. Concentrations in phytoplankton/detritus in the overlying water are calculated from the bioavailable water concentration C_W

$$C_A / C_W = L_A \cdot K_{OW} \qquad (5)$$

which assumes that C_A/C_W reflects the chemical's water-algae partition coefficient, which can be approximated by the chemicals K_{OW} and the lipid content of the organism L_A.

In applying the model to fish populations near pulp mills, whitefish were assumed to consume phytoplankton, and benthic and pelagic invertebrates, while suckers were assumed to accumulate TCDD and TCDF solely from benthic invertebrates. Water concentrations of TCDD/F were identical to those used for the Thomann model. Sediment concentrations (ng kg^{-1} dry weight) of TCDD and TCDF, and organic carbon levels, at each site were obtained by calculating geometric means of the data of Trudel (1991) and Mah et al. (1989). Lipid levels for fishes at each site were geometric means of monitoring data of Mah et al. (1989) and Whittle et al. (1990).

RESULTS AND DISCUSSION
Pharmacokinetic Studies

A wide range of assimilation efficiencies, depuration half-lives, and equilibrium BAFs were observed for six PCDD and nine PCDF congeners in juvenile rainbow trout (Table 1). Because exposure conditions (lighting, tank size, water flow temperature), fish size, and feeding rates were virtually identical the observed variation is due to characteristics of the individual congeners, especially the extent of chlorine substitution at the 2,3,7,8-positions.

Depuration of the fully 2,3,7,8-substituted PCDD/F's depuration could be described by single first-order decay curves and depuration half-lives were greater than 20 days. Within individual experiments coefficients of variation (CVs) of mean concentrations (generally N=3) in trout at each sampling time were generally less than 25%. Depuration of the non-2,3,7,8-substituted PCDDs, 1,2,3,7-TCDD, 1,3,6,8-TCDD, 1,2,3,4,7-PnCDD, as well as OCDD, was generally biphasic, with a rapid initial phase followed by a slower rate of elimination. Depuration half-lives for these compounds, calculated for the rapid phase where most of the compound was eliminated, were generally less than 5 days (Table 1). Depuration half-lives of non-2,3,7,8-substituted PCDFs were similar to those for the corresponding PCDDs (Table 1). Much longer half-lives and higher assimilation efficiencies were observed for 2,3,7,8-TCDF and 2,3,4,7,8-PnCDF under the same conditions. Where several dietary exposure concentrations of 2,3,7,8-TCDF and 2,3,4,7,8-PnCDF were studied, pharmacokinetic parameters were independent of exposure concentration over a 10- to 20-fold range (Table 1). This result reports the use of models based on first-order kinetics, e.g., Equation 1, 2, and 3, for describing bioaccumulation.

Assimilation efficiencies of the non-2,3,7,8-substituted PCDD/F's, as well as OCDD and OCDF, were generally < 20%. Although E is in principle a kinetic parameter that is independent of metabolic transformation rate, it is difficult to separate out the metabolic component without using radiolabelled compounds. Based on previous studies with ^{14}C-labelled dioxins the low values of E probably reflect metabolic and steric factors. Radiolabelled (^{14}C) 1,2,3,7-TCDD and 1,2,3,4,7-PnCDD were accumulated relatively efficiently from food by rainbow trout based on total ^{14}C in whole fish, but appeared to be rapidly transformed to products unextractable with toluene (Muir and Yarechewski 1988); therefore net assimilation, based on toluene extractable radioactivity,

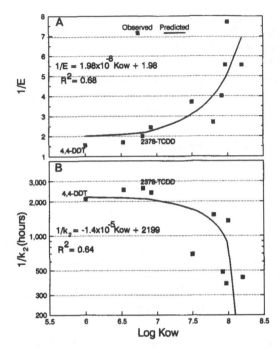

Figure 1   (A) Plot of the inverse of the assimilation efficiency (1/E), and (B) the inverse of the depuration rate ($1/k_2$) of 2,3,7,8-substituted PCDDs and PCDFs and 4,4'-DDT in juvenile rainbow trout. Individual PCDD/F congeners are listed in Table 1. Results for 2,3,7,8-TCDD were calculated from the data of Kleeman et al. (1986) for juvenile rainbow trout.

was low. Similar pathways probably explain the low assimilation of non-2,3,7,8-TCDFs(Table 1).

There was a significant (P < 0.05) linear relationship between the inverse of E and K_{OW} for 2,3,7,8-substituted PCDD/F congeners (Figure 1).

$$1/E = 3.12(\pm 0.76) \times 10^{-8} K_{ow} + 1.98(\pm 1.22)$$
$$(R = 0.68) \tag{6}$$

A similar relationship was obtained by Gobas (1988) for assimilation of a halogenated aromatics by fishes. Clark et al. (1990) found a 20-fold smaller value of a slope of Equation 6, which they termed a ratio of gut absorption aqueous resistance to egestion resistance, using only the data of Niimi and Oliver (1983) for assimilation of PCBs by large rainbow trout following a single oral capsule dosing. Clark et al. (1990) noted that these resistances are likely to vary with fish size.

The assimilation data also showed a good fit to the log E versus log K_{ow} relationship described by Thomann (1989) and Thomann et al. (1991, this volume) for compounds with log K_{ow} > 6.

$$\log E = 1.59(\pm 0.11) - 0.29(\pm 0.05) \cdot \log K_{ow}$$

$$\left(R^2 = 0.81\right) \tag{7}$$

Thomann (1989) and Opperhuizen and Sijm (1990) have noted that assimilation efficiencies decline above log K_{ow} = 6. The slope of the line is similar to the value of 0.25 reported by Thomann (1989) for the log E versus log K_{ow} relationship of a series of PCBs and TCDD.

There was a linear relationship (at P < 0.05, eight degrees of freedom) between the inverse of the depuration rate constant (1/k_2) and K_{ow} (Figure 1).

$$1/k_2 = -1.4(\pm 0.40) \cdot 10^{-5} K_{ow} + 2200(\pm 58)\text{h}$$

$$\left(R^2 = 0.64\right) \tag{8}$$

Also included in the data (Figure 1B) were results for 4,4-DDT (Muir and Yarechewski 1988) and for 2,3,7,8-TCDD calculated from Kleeman et al. (1986). Clark et al. (1990) using results for depuration of PCBs by rainbow trout from Niimi and Oliver (1983), obtained an expression with a similar intercept but with a positive slope. The greater hydrophobicity and molecular volume of PCDD/F's than correspondingly substituted PCBs may explain the negative value of the slope in Equation 6 for 2,3,7,8-substituted PCDD/Fs. The result suggest that as log K_{ow} increases above 7.5 resistance to elimination declines. This is a consequence of relatively rapid depuration rates for the high K_{ow} congeners, OCDD and OCDF, which may not be truly assimilated due to steric hindrance to absorption via the gut. Fish size and species may also be important. Values of the slope, which Clark et al. (1990) termed "gut absorption aqueous resistance" for PCBs and other halogenated aromatics, were in the 10^{-3} and 10^{-4} range, for goldfish and guppies, respectively. These fishes were of approximately the same size as those used in the present work, whereas the rainbow trout used in studies of PCB elimination (Niimi and Oliver 198) were about 10-fold heavier than those in our studies.

Mesocosm Studies

Whole water concentrations of 2,3,7,8-TCDF in mesocosms water declined rapidly from a range of 70 to 110 pg l^{-1} at 6 hr after addition of the chemical, to 20 pg l^{-1} after three days. After 21 days concentrations were below 5 pg l^{-1}. Average concentrations of TCDF in 16 sediment samples taken from day 3 to day

60 was 19 pg g^{-1}, with no apparent trend with time or significant differences between mesocosms. No degradation products of TCDF were detected in sediments extracts assayed by reverse-phase HPLC/LSC. From 80 to 90% of TCDF concentrations in the water column were associated with particles and the remainder was associated with dissolved organic carbon. Freely dissolved concentrations could not be measured after centrifugation and therefore must have represented less than 5% of whole water concentrations.

Concentrations of 2,3,7,8-TCDF in larval Chironomidae (aquatic midges), *Hexagenia* nymphs, and emerging insects (predominantly chironomids) reached maximum concentrations at 5 to 15 days after addition of TCDF (Figure 2). Chironomids collected from bottom sediments had TCDF concentrations ranging from 91 to 1320 pg g^{-1} lipid (wet) wt, with a mean concentration of 687 pg g^{-1} lipid wt (Figure 2). The emerging insects had TCDF concentrations ranging from 87 to 1310 pg g^{-1} lipid wt, with a mean concentration for all mesocosms of 990 pg g^{-1} lipid wt (Figure 2). *Hexagenia* had concentrations ranging from 48 to 432 pg g^{-1} lipid wt with maximum concentrations at 21 days after TCDF addition.

Values of E and k$_2$ for chironomids, *Hexagenia* and emerging insects were derived by fitting the accumulation data (Figure 2) for days 0 to 29 (emerging insects) and days 0 and 9 (others) to Equation 1 using iterative nonlinear regression with E and k$_2$ as unknown parameters. Water concentrations were average dissolved concentrations in sediment pore waters over the exposure period. Feeding rates of 0.1 and 0.3 g dry wt · g dry wt organism^{-1}·day $^{-1}$ for chironomids and *Hexagenia*, respectively, were assumed based on literature values of Davies (1975) and Dermott (1981). The best fit by iterative nonlinear regression was achieved for both emerging insects and chironomid larvae with E = 0.15 and k$_2$ = 0.02 d^{-1}. The analysis yielded lower values of E and larger K$_2$ (shorter half-lives) for TCDF in *Hexagenia* (Table 2).

Food Chain Modeling

The ability of the two food chain models to predict TCDD and TCDF concentrations in fish near bleached kraft mills was examined using data from Mah et al. (1989), and from the Canada Department of Fisheries and Oceans national Pulp Mill Dioxin Program (Whittle et al. 1990). In the latter reports the largest TCDD/F data base exists for whitefish and for various species of suckers, longnose sucker (*Catostomus catostomus*), largescale sucker (*Catostomus machrocheilus*), and white sucker (*Catostomus commersoni*). Composite samples comprising two to five fish were analysed for most sites. Median weights of the fish in the composites ranged from about 300 to 1300 g. For modeling purposes, i.e., for deriving fish volumes, and scaling uptake and depuration rates, a fish size of 500 g was used.

Pharmacokinetic parameters for 2,3,7,8-TCDF for emerging insects (Table 2) were used for both TCDD and TCDF in the invertebrate trophic level. For fish

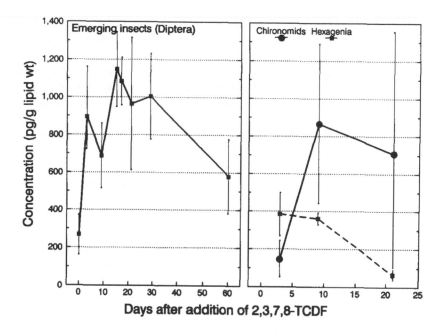

Figure 2. Concentrations (pg g^{-1} lipid wt) of 2,3,7,8-TCDF in emerging insects (mainly Diptera), and in sediment-dwelling chironomids and *Hexagenia*, from mesocosms treated with ^{3}H-2,3,7,8-TCDF sorbed on sediment particles. Data points represent means of 4 samples of emerging insects and of 5 to 15 individual analyses in the case of chironomids and *Hexagenia*.

Table 2. Estimated Bioaccumulation Parameters for 2,3,7,8-TCDF in Chironomid Larve and Emerging Insects from Lake Mecocosms

Invertebrate	Feeding rate[a] (g·g^{-1}·d^{-1})	Assimilation efficiency	Elimination half-life (days)
Chironomid	0.1	0.15	35
Emerging insects	0.1	0.15	35
Hexagenia	0.3	0.05	25

[a] Davies (1975) reported 0.1 to 0.5 g·g^{-1}·d^{-1} for Chironomids; Dermott, 1981 reported 0.3 to 0.5 g·g^{-1}·d^{-1} for Hexagenia. Both are based on g dry wt·g dry wt organism^{-1}·day^{-1}.

feeding on emerging insects values of E for TCDF and TCDD calculated for rainbow trout were used (Tables 1 and 3). Selection of depuration rates for TCDD/F was difficult because there is evidence that k_2 varies with fish size. Kuehl et al. (1986) reported half-lives for TCDD of 336 days in large (1 kg) carp transferred to the laboratory, and Cook et al. (1991) report a half-life of 61 days in smaller carp (15 g) following lab exposure. Thus the TCDF elimination data for small rainbow trout (Table 1) could not be used directly. The depuration rates reported by Kuehl et al. (1986) for TCDD and TCDF were selected because they are the best estimates for fish of similar size to those analyzed by the monitoring programs.

Feeding rates of 0.1 g dry wt/g dry wt organism/day were assumed for Diptera and Ephemeroptera based on literature values (Davies 1975), while feeding rates for large fish were assumed to be 1% of body weight (Thomann and Connolly 1984) in the absence of data for the fish near pulp mills. Growth rates of large fish and invertebrates were assumed to be zero in this steady-state version of the model. In the case of mountain whitefish, growth rates of adult fish (three to eight years) reported by Pettit and Wallace (1975), ranged from 7 to 10% per year, and thus would have relatively little impact on predicted concentrations in fish.

No pharmacokinetic parameters were required for the Gobas model. However, the model-generated uptake and depuration rate constants and assimilation efficiencies from semi-empirical relationships with K_{ow} and fish weights (Gobas 1991). The k_1 values for fish generated by the model were similar to those used for the Thomann model, while k_2 values for TCDD and TCDF for the two models differed by less than a factor of two (Table 3). Thus the ratios k_1/k_2 (equal to the equilibrium BCFs) used in both models were similar.

TCDD and TCDF concentrations (on a lipid wt basis) predicted by the two models are compared in Figures 3 and 4 and in Table 4, with observed concentrations in fish sampled during 1988 to 1990 in surveys near chlorine bleached kraft mills in British Columbia (Mah et al. 1989), and five other Canadian provinces (Whittle et al. 1990). These sites were selected because data on TCDD and TCDF concentrations were available for mill effluents (PPRIC 1989, 1991), bed sediments (Trudel 1991, Mah et al. 1989) and fishes, along with dilution factors based on average annual river flows, sediment organic carbon, and fish lipid content. Coastal mills and those located on large lakes were excluded because of the difficulty of estimating dilution factors. TCDF was detectable in mill effluents and bed sediments at almost all locations while TCDD was generally below detection limits (< 2 ng kg^{-1} in sediment and < 10 ng l^{-1} in effluents), especially in sediment. In effluents, TCDD concentrations were assumed to be equal to one tenth those of TCDF, the approximate TCDD/TCDF ratio at B.C. mills in 1988 (PPRIC 1989), while in sediments the detection limit was used as the TCDD level.

Both models gave good predictions of lipid weight concentrations of TCDD and TCDF in whitefish at seven of eight locations (Figure 3). For TCDD the

Table 3.   Parameters used for Thomann and Gobas Food Chain Models

Parameter	Compartment	Compound	Value	Reference
Log K_{ow}		TCDD	6.8	Shiu et al. (1988)
		TCDF	6.5	Sijm et al. (1989)

Thomann Model Parameters

Parameter	Compartment	Compound	Value	Reference
k_1	fish	TCDD	108 $g \cdot ml^{-1} \cdot d^{-1}$	Cook et al. (1990)
		TCDF	108 $g \cdot ml^{-1} \cdot d^{-1}$	[a]
k_1	invertebrates	TCDD	1080 $g \cdot ml^{-1} \cdot d^{-1}$	[b]
		TCDF	1080 $g \cdot ml^{-1} \cdot d^{-1}$	[a]
k_2	fish	TCDD	0.002 d^{-1}	Kuehl et al. (1986)
		TCDF	0.003 d^{-1}	Kuehl et al. (1986)
	invertebrates	TCDD	0.02 d^{-1}	[c]
		TCDF	0.02 d^{-1}	This study
E	fish	TCDD	0.50	Kleeman et al. (1986)
		TCDF	0.55	This study
	invertebrates	TCDD	0.15	[c]
		TCDF	0.15	This study
F[d]	fish		0.009 $g \cdot g^{-1} \cdot d^{-1}$	Thomann (1981)
	invertebrates		0.10 $g \cdot g^{-1} \cdot d^{-1}$	Davies (1975)

Gobas Model Parameters

Parameter	Compartment	Compound	Value	Reference
Volume	fish		0.5 l	[e]
Density	benthos		1 kg l^{-1}	
	sediment		1.5 kg l^{-1}	
Lipid	fish		0.005-0.10	[f]
	invertebrates		0.02	
Particulate organic carbon in water			3×10^{-6}	
k_1	suckers and whitefish	(TCDD/F)	117	[g]
k_2	suckers	TCDD	0.003-0.005 d^{-1}	[g]
	whitefish	TCDD	0.003-0.005 d^{-1}	[g]
	suckers	TCDF	0.004-0.005 d^{-1}	[g]
	whitefish	TCDF	0.004-0.008 d^{-1}	[g]

Table 3. (continued)

a Assumed to be equal to TCDD uptake rate reported for lake trout by Cook et al. (1990). Based on tissue wet weight.

b Estimated to be 10× the value for fish (Thomann 1981).

c Assumed to be the same as the values for TCDF.

d Feeding rate expressed as g food·g fish⁻¹d⁻¹ or as fraction of body volume.

e Approximate median size of fish sampled by Mah et al. (1989) and the Fisheries and Oceans National Pulp Mill Dioxin program (Whittle et al. 1990).

f Range of geometric mean lipid from all locations. Actual means from each site were used.

g Calculated from semi-empirical relationships built into the model of Gobas (1991).

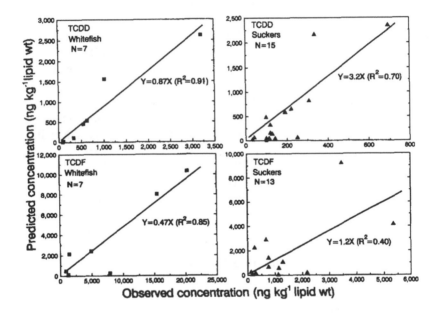

Figure 3. Plots of 2,3,7,8-TCDD and -TCDF concentrations, predicted with the "generic" Thomann model, and observed concentrations (lipid weight basis) of 2,3,7,8-TCDD and -TCDF in whitefish (*Prosopium williamsoni*) and suckers (*Catostomus* sp) collected near inland bleach kraft mills in Canada. Locations are given in Table 4. The model assumed dilution and partitioning of TCDD/F in mill effluents to particles, and hence to filter-feeding invertebrates, was the major pathway to fishes. Lines represent linear regressions of predicted versus observed concentrations (intercept 0).

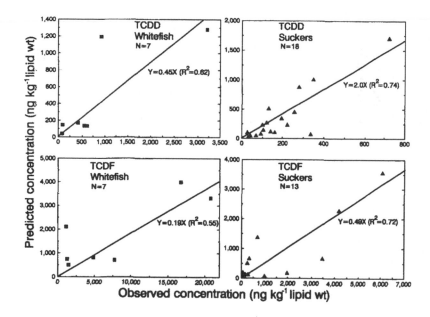

Figure 4. Plots of 2,3,7,8-TCDD and -TCDF concentrations, predicted with the model of Gobas et al. (1991), and observed concentrations (lipid weight basis) of 2,3,7,8-TCDD and -TCDF in whitefish and suckers. Most locations (Table 4) are the same as those in Figure 3. The model assumed that TCDD/F in bed sediments were in equilibrium with benthic organisms and that both fish species were contaminated mainly via benthos. Lines represent linear regressions of predicted versus observed concentrations (intercept 0).

outlier was a composite sample of seven fish from Quesnel on the Fraser River (B.C.), which had far higher levels than predicted on the basis of effluent concentrations at the bleached kraft mill. Because Quesnel is downstream from Prince George, and mountain whitefish were collected during their fall migratory period, the possibility exists that these fish were contaminated by the upstream source. Omitting this outlier, linear correlation (intercept 0) of predicted TCDD concentrations in whitefish versus observed results for both models were highly significant ($R^2 = 0.91$ and 0.62 for the Thomann and Gobas models, respectively). The Thomann model also predicted TCDF concentrations in whitefish for the same seven locations ($R^2 = 0.85$). Slopes of the predicted versus observed whitefish data for the kinetic model approached 1.0, indicated good predictive ability for TCDD/F, based on dilution of mill effluent and partitioning to particulates in the water column, without considering sediment concentrations.

Table 4. Observed Concentrations of 2,3,7,8-TCDD and TCDF in Fishes (Geometric Means Lipid Wt) and Concentrations Predicted with the Thomann and Gobas Food Chain Models

Location	Tissue[d]	Observed (µg kg⁻¹)				Predicted-Thomann[b] (µg kg⁻¹)				Predicted-Gobas[c] (µg kg⁻¹)			
		TCDD		TCDF		TCDD		TCDF		TCDD		TCDF	
		WF	S	WF	S	WF	S	WF	S	WF	S	WF	S
Columbia, R., Castlegar, BC	M	0.06-0.12	<0.10	6.2-10	1.3	0.01	0.04	0.26	1.1	0.04	0.22	0.71	4.0
Kootenay R., Cranbrook, BC	M	<0.10	<0.20-<0.33	1.3	3.9-4.5	0.03	0.05	0.11	0.23	0.15	0.45	0.75	2.3
Fraser R., Kamloops, BC	M	3.2	0.69	20	0.89	2.6	2.3	10	9.2	1.3	1.7	90	67
Fraser R., Prince George, BC	M	0.13-6.5	<0.13-0.73	2.9-97	2.9-9.8	1.6	0.82	8.0	4.1	1.2	1.0	4.0	3.6
Fraser R., Quesnel, BC	M	2.4	0.12-2.4	21	0.63-10	0.01	0.03	0.06	0.11	0.35	0.87	3.3	8.0
Wapiti R., Grand Prairie, AL	M+WF	0.56-0.70	0.16-0.34	0.54-0.86	3.6-6.5	0.54	0.65	2.4	2.9	0.13	0.23	0.81	1.4
Athabasca R., Hinton, AL	M+WF	0.30-1.7	0.08-0.12	0.94-4.0	0.13-0.44	0.45	0.48	2.2	2.2	0.14	0.14	0.50	0.49
Saskatchewan, R., Pr. Albert, SK	WF	-[e]	0.04	-	0.12	-	0.07	-	0.25	-	0.04	-	0.05
Wabigon R./Clay L., Dryden, ON	M	0.30-0.61	-	0.90-1.5	-	0.13	-	0.57	-	0.17	-	2.1	-

Location	Tissue[d]												
Kaministikwia R., Thunder Bay, ON	M	-	0.02-0.03	-	0.06-0.09	-	0.32	-	1.4	-	0.09	-	0.18
Spanish R., Espanola, ON	M	-	0.09-0.21	-	0.74-1.7	-	0.13	-	0.56	-	0.50	-	5.3
St. Maurice R., LaTuque, QU	WF	-	<0.01-0.96	-	0.04-10	-	2.5	-	9.2	-	0.06	-	0.65
Ottawa R., Portagedu-Fort, QU	M+WF	-	<0.08-0.25	-	0.13-0.61	-	0.04	-	0.17	-	0.11	-	0.11
Bell R., Quevillon, QU	M+WF	-	0.11-0.34	-	0.29-0.46	-	-[f]	-	-	-	0.34	-	0.65
Mistassini R., St. Felician, QU	M+WF	-	0.06-0.07	-	1.8-1.9	-	-[f]	-	-	-	0.04	-	0.12
Ottawa R., Thurso, QU	WF	-	<0.07-0.16	-	0.98-1.5	-	0.01	-	0.03	-	0.10	-	0.05
St. Francois R., Windsor, QU	WF	-	<0.09-0.19	-	2.0-3.3	-	0.04	-	0.18	-	0.07	-	0.17
St. John R., Nackawic, NB	M+WF	-	0.01-0.05	-	0.09-0.39	-	0.02	-	0.06	-	0.02	-	0.11
Mirimachi R., Newcastle, NB	M+WF	-	0.10-0.21	-	0.59-1.3	-	0.15	-	0.65	-	0.26	-	3.0

[a] Range observed lipid weight concentrations in whitefish (WF) or sucker (S) muscle and/or whole fish from Mah et al. (1989) and from the Dept. of Fisheries and Oceans National Pulp Mill Dioxin Program (Whittle et al. 1990).

[b] Concentrations on a lipid weight basis predicted with a "generic" Thomann model linked to mill effluent concentrations and not to bed sediments.

[c] Concentrations on a lipid weight basis predicted with the model of Gobas et al. (1991) which included estimated water concentrations and bed sediment concentrations (from Trudel 1991).

[d] Sample tissue analysed: muscle = M; whole fish = WF; where both samples were analysed = M+WF.

[e] Dash indicates no data available for this species.

[f] Not determined because of lack of data on mean annual river flow. For the Gobas model, accumulation was assumed to be entirely from bed sediments.

The Gobas model adequately predicted TCDF levels observed in whitefish at Quesnel but overestimated concentrations in whitefish at Kamloops. The Kamloops site was characterized by unusually high TCDF concentrations in bed sediment (Table 4) and this resulted in high predicted levels because of the model's assumption of bed sediment-benthos equilibrium. With Kamloops omitted, concentrations predicted by the Gobas model were significantly correlated with observations ($R^2 = 0.55$) although levels were generally underestimated (slope $= 0.19 \pm 0.03$).

Concentrations of TCDD were available for suckers at 18 sites near Canadian bleached kraft mills, and both models successfully predict the observed levels at most locations (Figures 3 and 4). Predicted concentrations of TCDD were significantly correlated with observed results for both models but levels were overestimated (slope $= 3.2 \pm 0.43$; $R^2 = 0.70$ for the kinetics-based model and (slope $= 2.0 \pm 0.22$; $R^2 = 0.74$ for the Gobas model). Because only mill effluent levels of TCDD were used in the kinetics-based model, two sites in northern Quebec (Mistassini River, St. Felicien, and Bell River, Lebel-dur-Quevillon) could not be included because average annual flow data were unavailable. Data for Quesnel were not used for the reasons described in the case of whitefish. All 18 sites were used for the Gobas model predictions because data on TCDD/F in sediment were available. The overestimation of TCDD levels in suckers is likely related to the use of detection limits, or in some cases fractions of observed TCDF levels, to estimate concentrations of TCDD in sediment and mill effluent.

The models were less successful in predicting TCDF concentrations in suckers from the same 18 locations. A significant correlation between predicted and observed results for the Thomann model (slope $= 1.2 \pm 0.28$; $R^2 = 0.40$) could be obtained only if data for Kamloops and Cranbrook were omitted, along with the three other sites mentioned previously. The model overestimated TCDF levels in suckers at Cranbrook and underestimated them at Kamloops. Cranbrook was the only site for which TCDF was undetectable in mill effluents in both 1988 and 1990 surveys (PPRIC 1989, 1991). The Gobas model greatly overestimated observed TCDF levels at five of 18 sites. In all cases, elevated levels in sediment at these sites did not correspond well to TCDF levels in suckers. With these sites removed predicted and observed levels at 13 locations were significant correlated (slope $= 0.49 \pm 0.07$; $R^2 = 0.72$).

The lack of agreement between predicted and observed levels at some locations is not surprising given the nature of the monitoring data for sediment, mill effluents, and fishes. Sediments collected from sites downstream of the mills generally consisted of sand or sandy silt and with a wide range of organic carbon (0.2 to 13%) (Trudel 1991). The observed TCDF concentrations in bed sediments may not be representative of levels on organic-rich particles consumed by invertebrates filtering water or ingesting sediment. For modeling purposes concentrations on suspended particles would be more useful than bottom material collected with an Ekman dredge, however, this data is available for only a limited number of sites. Merriman (1988) detected 11.5 ng kg^{-1} (dry wt) of

TCDD (but not TCDF) in suspended sediments collected by continuous centrifugation from the Rainy River downstream of bleached kraft mills.

The exact locations of the fish populations sampled by Mah et al. (1989) and Whittle et al. (1990), and hence the duration of their exposure to pulp mill effluent discharges are unknown. However, exposure is probably not continuous because both species undergo annual migrations. Suckers undergo early spring spawning migrations to smaller tributaries (Scott and Crossman 1973), while mountain whitefish migrate upstream in late spring and summer (Thompson and Davies 1976). After spawning in the fall, mounting whitefish return downstream to overwintering areas. Both species overwinter in large rivers where they may be exposed to pulp mill effluents under low flow conditions.

Both models tended to underestimate TCDF concentrations in whitefish. The Gobas model also underestimated TCDF in suckers. This reflects the difficulty in simulating the feeding preferences of both species. Mountain whitefish are visual feeders feeding primarily on drifting Diptera, Ephemeroptera nymphs, and Trichoptera (Bond and Berry 1980, Thompson and Davies 1976). Suckers also feed on benthic invertebrates, especially Hemiptera and Pelecypoda. Detritus also accounts for a significant portion of stomach contents of suckers due to their bottom grazing behavior (Bond and Berry 1980). Mountain whitefish, by their tendency to feed on emerging insects and filter feeders in the water column, may therefore feed at a higher trophic level than suckers. Further evidence for this is the higher concentrations of TCDD and TCDF generally observed in whitefish at sites where both species were located (Mah et al. 1989, Whittle et al. 1990). Species differences in lipid-based bioaccumulation parameters are less likely to explain lower concentrations in suckers than in whitefish. There are not large species differences between rainbow trout, fathead minnows, and carp in lipid-based bioaccumulation parameters for TCDD (Cook et al. 1991).

Apart from analyses of pooled samples of benthic insects near Hinton and Grand Prairie (Noton 1991) in which had TCDF ranged from 20 to 217 ng kg^{-1} and TCDD from 5 to 17 ng kg^{-1} we found no data with which to verify predictions for TCDD/F in invertebrates. The concentrations from the two Alberta sites were within the range predicted for invertebrates by both models.

In conclusion, despite several simplifying assumptions (steady-state concentrations and growth, rapid equilibration of effluent with particles, three trophic levels, mean annual river flow) both food chain models were able to predict TCDD/F concentrations in whitefish and suckers at the majority of locations. Perhaps the greatest simplification was the use of a mean annual flow for dilution of mill effluent. River flows at all locations vary seasonally with high flows in the April to June period. Use of winter low flow dilutions would have given about 3- to 10-fold higher predicted TCDD/F concentrations in the dissolved phase and in detritus and food chain organisms with the Thomann model. The lack of data on temporal trends in TCDD/F levels in fish, invertebrates, or sediments near the mills precluded a more detailed analysis. Modeling with a single mean concentration in water and particles seems best suited for food chain accumulation of persistent compounds like TCDD and TCDF in large fish because changes in levels of TCDD/F will occur slowly due to slow rates of depuration and growth.

A future test of these modeling approaches will be the prediction of TCDD and TCDF concentrations where levels in effluents are below detection limits, i.e., probably <1 pg⁻¹ for both congeners. The effluent and fish data used here are based on surveys in 1988 to early 1990 when changes to reduce chlorine use in the mills had just begun to be implemented. If TCDD/F are nondetectable in effluents the major source of contaminants to fish will be the existing loading in bottom sediments. More accurate predictions of levels in fish will depend on knowledge of the fate of contaminants on bed and suspended sediments and use of nonsteady-state solutions for food chain modeling. This will require a more sophisticated approach to generate exposure concentrations such as that of Connolly and Tonelli (1985), which takes into account the dynamics of river hydrology, erosion of bed sediments and suspended particles, and diffusion from sediment pore waters. Validation of the approach will require frequent sampling of key parameters such as TCDD/F concentrations in bed solids and suspended sediment and food chain organisms.

ACKNOWLEDGMENTS

Funding for studies of TCDF pharmacokinetics and bioavailability was provided by a Natural Sciences and Engineering Council of Canada Strategic Grant. We thank Dr. Frank Gobas for providing the food chain model and for helpful discussion of modeling approaches. Thanks are also due to Peter Delorme and Scott Brown for reviewing the manuscript.

REFERENCES

Amendola, G.A., R.E. Handy and D.G. Bodien. "Bench Scale Study of Dioxins and Furan (2,3,7,8-TCDD and 2,3,7,8-TCDF) Treatability in Pulp and Paper Mill Wastewaters," *TAPPI Journal* (1989) pp.189-195.

Bond, W.A., and D.K. Berry. "Fishery Resources of the Athabasca River Downstream of Fort McMurray, Alberta. Vol. II," Alberta Oil Sands Environmental Research Program, Alberta Department of Environment, Edmonton (1980) p. 158.

Bruggeman, W.A., L.B.J.M. Martron, D. Kooiman and O. Hutzinger, "Accumulation and Elimination Kinetics of Di- , Tri- and Tetrachlorobiphenyls by Goldfish after Dietary and Aqueous Exposure," *Chemosphere* 10:811-832 (1981).

Clark, K.E., F.A.P.C. Gobas and D. Mackay. "Model of Organic Chemical Uptake and Clearance by Fish from Food and Water," *Environ. Sci. Technol.* 24:1203-1213 (1990).

Clement, R.E., S.A. Suter, E. Reiner, D. McCurvin and D. Hollinger. "Concentrations of Chlorinated Dibenzo-p-dioxins and Dibenzofurans in Effluents and Centrifuged Particulates from Ontario Pulp and Paper Mills," *Chemosphere* 19:649-654 (1989).

Cook, P.M., A.R. Batterman, B.C. Butterworth, K.B. Lodge and S.W. Kohlbry. "Laboratory Study of TCDD Bioaccumulation by Lake Trout from Lake Ontario Sediments, Food Chain and Water," in *Lake Ontario TCDD Bioaccumulation Study,* Final Report, U.S. EPA Duluth, WI (1990).

Cook, P.M., M.K. Walker, D.W. Kuehl and R.E. Peterson. "Bioaccumulation and Toxicity of 2,3,7,8-Tetrachlorodibenzo-*p*-dioxins and Related Compounds in Aquatic Ecosystems," in *Banbury Report: Biologic Basis for Risk Assessment of Dioxins and Related Compounds*, Vol. 35, M.A. Gallo, R.J. Scheuplein and C.A. Vande Heijde, Eds., Cold Spring Harbor Laboratory Press, Cold Spring Harbor, NY, pp. 143-167 (1991).

Connolly, J.P., and C.J. Pedersen. "A Thermodynamic-based Evaluation of Organic Chemical Accumulation in Aquatic Organisms," *Environ. Sci. Technol.* 22:99-103 (1989).

Connolly, J.P., and R. Tonelli. "Modelling Kepone in the Striped Bass Food Chain of the James River Estuary," *Estuar. Coast. Shelf Sci.* 20:349-366 (1985).

Davies, I.J. "Selective Feeding in some Arctic Chironomidae," *Verh. Internat. Verein. Limnol.* 19:3149-3154 (1975).

Dermott, R. "Ingestion Rate of the Burrowing Mayfly *Hexagenia limbata* as determined with ^{14}C," *Hydrobiologia* 83:499-503 (1981).

Endicott, D., W.R. Richardson and D.M. DiToro. "Lake Ontario TCDD Modeling Report," in *Lake Ontario TCDD Bioaccumulation Study*, Final Report, U.S. EPA, Duluth, WI (1990).

Fairchild, W.L., D.C.G. Muir, R. Currie, and A.L. Yarechewski. Emerging Insects as a Biotic Pathway for Movement of 2,3,7,8, - Tetrachlorodibenzofuran From Lake Sediments. *Environ. Toxicol. Chem*, 11:867-872 (1992)

Gobas, F.A.P.C., D.C.G. Muir and D. Mackay."Dynamics of Dietary Bioaccumulation and Faecal Elimination of Hydrophobic Organic Chemicals in Fish," *Chemosphere* 17:943-962 (1988).

Gobas, F.A.P.C. "Modelling the Accumulation and Toxicity of Organic Chemicals in Aquatic Food-chains," in *Chemical Dynamics in Freshwater Ecosystems*, F.A.C.P. Gobas and F. McCorquadale, Eds., (Ann Arbor, MI: Lewis Publishers, 1991).

Haliburton, D. Environment Canada, Ottawa K1A 0H3, Personal communication (1990).

Harding, L.E., and W.M. Pomeroy. "Dioxin and Furan Levels in Sediments, Fish and Invertebrates from Fishery Closure Areas of Coastal British Columbia," Regional Data Report, Environment Canada, Environmental Protection, Pacific and Yukon Region, North Vancouver, B.C. (1990), pp. 67 and Appendix.

Herbes, S.E., and C.P. Allen. "Lipid Quantification of Freshwater Invertebrates: Method Modification for Microquantitation," *Can. J. Fish. Aquat. Sci.* 40:1315-1317 (1983).

Karickhoff, S.W. "Semi-empirical Estimation of Sorption of Hydrophobic Pollutants on Natural Sediments and Soils,"*Chemosphere* 10:833-846 (1979).

Kleeman, J.M., J.R. Olson, S.M. Chen and R.E. Peterson. "Metabolism and Disposition of 2,3,7,8-tetrachlorodibenzo-*p*-dioxin in Rainbow Trout," *Toxicol. Appl. Pharmacol.* 83:391-401 (1986).

Kuehl, D.W., P.M. Cook and A.R. Batterman. "Uptake and Depuration Studies of PCDDs and PCDFs in Freshwater Fish," *Chemosphere* 15:2023-2026 (1986).

Kuehl, D.W., P.M. Cook, A.R. Batterman, D. Lothenbach and B.C. Butterworth. "Bioavailability of Polychlorinated Dibenzo-*p*-dioxins and Dibenzofurans from Contaminated Wisconsin River Sediment to Carp," *Chemosphere* 16:667-679 (1987).

Kuehl, D.W., B.C. Butterworth, A. McBride, S. Kroner and D. Bahnick. "Contamination of Fish by 2,3,7,8-Tetrachlorodibenzo-*p*-dioxin: a Survey of Fish from Major Watersheds in the United States," *Chemosphere* 18:1997-2014 (1989).

Mah, F.T.S., D.D. MacDonald, S.W. Sheehan, T.M. Touminen and D. Valiela. "Dioxins and Furans in Sediment and Fish from the Vicinity of Ten Inland Pulp Mills in British Columbia," Water Quality Branch, Environment Canada. Vancouver, B.C. (1989), p. 77.

Merriman, J.C. "Distribution of Organic Contaminants in Water and Suspended Solids of the Rainy River," *Water Pollut. Res. J. Canada* 23:590-600 (1988).

Muir, D.C.G., and A.L. Yarechewski. "Dietary Accumulation of Four Chlorinated Dioxin Congeners by Rainbow Trout and Fathead Minnows, *Environ. Toxicol. Chem.* 7:227-236 (1988).

Muir, D.C.G., A.L. Yarechewski, D.A. Metner, W.L. Lockhart, G.R.B. Webster and K.J. Friesen. "Dietary Accumulation and Sustained Hepatic Mixed Function Oxidase Enzyme Induction by 2,3,4,7,9-Pentachlorodibenzofuran in Rainbow Trout," *Environ. Toxicol. Chem.* 9:1465-1474 (1990).

Niimi, A.J., and B.G. Oliver. "Biological Half-Lives of Polychlorinated Biphenyl (PCB) Congeners in Whole Fish and Muscle of Rainbow Trout (*Salmo Gairdneri*)," *Can. J. Fish. Aquat. Sci.* 40:1388-1394 (1983).

Norstrom, R.J., A.E. McKinnon and A.S.W. deFreitas. "A Bioenergetics-Based Model for Pollutant Accumulation by Fish. Simulation of PCB and Methylmercury Residue Levels in Ottawa River Yellow Perch (*Perca Flavescens*)," *J. Fish. Res. Board Can.* 33:248-267 (1976).

Noton, L. Alberta Environment, Environmental Quality Monitoring Branch, Edmonton, Alta. Personal communication (1991).

Opperhuizen, A., and D.T.H.M. Sijm. "Bioaccumulation and Biotransformation of Poloychlorinated Dibenzo-*p*-dioxins and Dibenzofurans in Fish," *Environ. Toxicol. Chem.* 9:175-186 (1990).

Pettit, S.W., and R.L. Wallace. "Age, Growth and Movement of Mountain Whitefish (*Prosopium Williamsoni*) (Girard) in the North Fork Clearwater River, Idaho," *Trans. Am. Fish. Soc.* 104:68-76 (1975).

PPRIC. "National Dioxin Survey 1989," Pulp and Paper Research Institute of Canada, Pointe Claire, QU (1989).

PPRIC. "National Dioxin Survey 1990," Pulp and Paper Research Institute of Canada, Pointe Claire, QU (1991).

Rogers, I.H., C.D. Levings, W.L. Lockhart and R.J. Norstrom. "Observations on Overwintering Juvenile Chinook Salmon (*Oncorhynchus Tshawytscha*) Exposed to Bleached Kraft Mill Effluent in the Upper Fraser River, British Columbia," *Chemosphere* 19:1853-1868 (1989).

Rubenstein, N.I., R.J. Pruell, B.K. Taplin, J.A. LiVolsi and C.B. Norwood. "Bioavailability of 2,3,7,8-TCDD, 2,3,7,8-TCDF and PCBs to Marine Benthos from Passaic River Sediments," *Chemosphere* 20:1097-1102 (1990).

Scott, W.B., and E.J. Crossman. "Freshwater Fishes of Canada," *Fish Res. Board Can.*, Bulletin 184, Department of Fisheries and Oceans, Ottawa Canada (1973).

Servos, M.R., D.C.G. Muir and G.R.B. Webster. "Effect of Particulate and Dissolved Organic Carbon on the Bioavailability of Polychlorinated Dioxins," *Aquatic. Toxicol.* 14:169-184 (1989).

Shiu, W.Y., W. Doucette, F.A.P.C. Gobas, A. Andren and D. Mackay. "Physical-chemical Properties of Chlorinated Dibenzo-*p*-dioxins," *Environ. Sci. Technol.* 22:651-658 (1988).

Sijm, D.T.H.M., H. Wever, P.J. de Vries and A. Opperhuizen. "Octan-1-ol/water Partition Coefficients of Polychlorinated Dibenzo-*p*-dioxins and Dibenzofurans: Experimental Values Determined with a Stirring Method," *Chemosphere* 19:263-266 (1989).

Thomann, R.V. "Equilibrium Model of Fate of Microcontaminants in Diverse Aquatic Food Chains," *Can. J. Fish. Aquat. Sci.* 38:280-296 (1981).

Thomann, R. V. "Bioaccumulation Model of Organic Chemical Distribution in Aquatic Food Chains," *Environ. Sci. Technol.* 23:699-707 (1989).

Thomann, R.V., and J.P. Connolly. "Model of PCB in the Lake Michigan Lake Trout Food Chain," *Environ. Sci. Technol.* 18:65-71 (1984).

Thomann, R.V., J.P. Connolly and T. Parkerton. "Modeling Accumulation of Organic Chemicals in Aquatic Food Webs," in *Chemical Dynamics in Freshwater Ecosystems*, F.A.P.C. Gobas and F. McCorquadale, Eds. (Ann Arbor, MI: Lewis Publishers, 1991).

Thompson, G.E., and R.W. Davies. "Observations on the Age, Growth, Reproduction and Feeding of Mountain Whitefish (*Prosopium Williamsoni*) in the Sheep River Alberta," *Trans. Am. Fish. Soc.* 105:208-219 (1976).

Trudel, L. "Dioxins and Furans in Bottom Sediments Near the 47 Canadian Pulp and Paper Mills Using Chlorine Bleaching," Water Quality Branch, Environment Canada, Ottawa, (1991), p. 70 and Appendix.

van der Weiden, M.E.J., L.H.J. Craane, E.H.G. Evers, R.M.M. Kooke, K. Olie, W. Seinen and M. van den Berg. "Bioavailability of PCDDs and PCDFs from Bottom Sediments and some Associated Biological and Toxicological Effects in the Carp *Cyprinus Carpio,*" *Chemosphere* 19:1009-1016 (1989).

Whittle, D.M., D.B. Sergeant, S. Huestis and W.H. Hyatt. "The Occurrence of Dioxin and Furan Isomers in Fish and Shellfish Collected near Bleached Kraft Mills in Canada," Paper presented at the 10th International Symposium on Chlorinated Dioxins and Related Compounds, Bayreuth, Germany, Sept. 1990.

9 TEMPORAL TRENDS AND DISTRIBUTION OF PCB CONGENERS IN A SMALL CONTAMINATED LAKE IN ONTARIO CANADA

INTRODUCTION

Lake Clear, a moderately small, isolated lake in eastern Ontario, Canada, was contaminated with polychlorinated biphenyls (PCBs) in the mid-1970s when PCB-contaminated oil was used as a dust suppressant on a road adjacent to the western end of the lake. Samples collected in 1986 and 1987 show that PCB levels in the biota have declined markedly since the early 1980s. The consistency of the pattern of PCB congeners in sediment profiles, and throughout all classes of biota indicated that there has been little "weathering" of PCBs with time. Trends in the distribution of individual PCB congeners throughout the lake ecosystem were tested statistically using two methods. ANOVA and range tests showed that the proportion of lower chlorinated congeners was significantly higher in the water and suspended solid phases than in sediments and biota. Principal component analysis indicated that the distribution of PCB congeners in the bottom sediments differed little from the Aroclor 1254 which originally contaminated the lake. The pattern of PCB congeners in biota were the same for all species; including short-lived or young organisms (e.g., zooplankton, young-of-the-year perch) and fish species greater than ten years of age (e.g., whitefish and lake trout). All of these analyses indicate that, while the PCBs available for uptake by biota are declining with time in the lake, the various congeners are being lost from circulation at equal rates. It is suggested that Lake Clear may be a simple model for predicting the fate and distribution of PCBs in large contaminated lake systems, such as the Great Lakes.

ISBN 0-87371-511-X

Background

The production of polychlorinated biphenyls (PCBs) in North America reached a peak in the late 1960s, but recognition of widespread environmental contamination by these compounds led to reduced production throughout the 1970s and a total ban on production in 1977. In some aquatic environments, such as the Great Lakes (International Joint Commission 1989, DeVault 1985, Suns et al. 1985), the level of PCB residues appears to have declined significantly in the past twenty years (Brown et al. 1985), but in other areas, PCB residues have remained constant (Norstrom et al. 1988, Addison et al. 1984, and Schmitt et al. 1985). It has been suggested that the decline in PCB residues in some aquatic systems may be attributed primarily to "weathering"; that is, a decline in the proportion of less chlorinated PCB congeners (Brown et al. 1987). Thus, moderately and higher chlorinated PCBs, which include the most toxic congeners (Kannan et al. 1988), may remain at constant levels in the aquatic environment because they are less volatile, more soluble in lipids, adsorb readily to sediments, and are more resistant to metabolic and microbial degradation (Kenaga and Goring 1980, Shiu and Mackay 1986, Murphy et al. 1987, Sawhney 1987, Connell 1988).

Both sedimentation and volatilization are important physical mechanisms for removal of PCBs from aquatic systems. Swackhamer and Armstrong (1986) determined that volatilization was as important a mechanism as sedimentation in the removal of PCBs from Lake Michigan. Similarly, volatilization was equal to or greater than sedimentation in removing PCBs from the water-column of a small remote lake (Swackhamer et al. 1988). In this latter study, losses through volatilization and sedimentation were greatest for less chlorinated congeners. Theoretical simulations of the partitioning of PCBs between water, particulates, and the atmosphere (Burkhard et al. 1985b) indicate that PCB mixtures may or may not weather with time, depending upon environmental conditions. Studies on microbial decomposition of PCBs in aquatic sediments (Brown et al. 1987, Quensen et al. 1988) indicate that there can be dechlorination and eventual aerobic degradation of PCBs in highly contaminated sediments. Less chlorinated congeners unsubstituted at the meta- and para- positions are most susceptible to dechlorination.

Very few studies have been done to document variations in the congener composition of PCBs in the various species and trophic levels of contaminated lakes. Oliver and Niimi (1988) showed that in Lake Ontario biota, there was a large difference in congener composition between the water phase and phytoplankton, but in higher trophic levels the congener composition remained relatively constant. Hence, it is expected that any weathering of PCBs would first become evident among the biotic community in short-lived organisms that accumulate PCBs by absorption from the aqueous phase (e.g., zooplankton).

It is difficult to test hypotheses regarding the decline in PCB residues using field data gathered from large aquatic ecosystems, primarily because of the large number of factors that influence the distribution of individual congeners. These interferences include rates of water replacement, continuing inputs from point sources and the atmosphere, sediment resuspension and mixing within the lake, and, in the case of biota, a range of factors including diet, age, and migratory behavior of individual species. In addition, large contaminated lakes often have a history of multiple point sources involving a number of different commercial PCB mixtures (i.e., Aroclors 1242, 1248, 1254, and 1260). These problems of data interpretation are evident in Lake Ontario, where PCB residues in fish continue to decline in some areas, but have recently begun to increase in other areas (Suns et al. 1983, Suns et al. 1985).

It is the purpose of this study to document historical trends in the decline of PCBs and to characterize the recent distribution of 19 PCB congeners in the biota, water, and sediments of Lake Clear, Renfrew County, Ontario, Canada; a moderately small freshwater lake contaminated with PCBs in the mid-1970s. We wish to determine whether the distribution of PCBs in the biota and sediments of contaminated lake systems are modified with time according to the physical properties and structure of the congeners. From the literature, it is expected that PCB residues will reach the highest concentrations in biota with the highest lipid levels, and these residues will decline slowly with time as older individuals are replaced by younger individuals. For the individual PCB congeners, it is predicted that there will be a relatively low proportion of less chlorinated congeners throughout the lake because of physical processes (sedimentation, volatilization) and microbial degradation. This "weathered" composition of the PCBs will be particularly evident in short-lived organisms within lower trophic levels. Using historical data, we will estimate the rate at which biota in the lake are recovering from the very high PCB concentrations reported in the late 1970s and early 1980s.

Materials and Methods
Study Area and Historical Data

Lake Clear is a moderately small, slightly eutrophic lake in the eastern part of Ontario, Canada (Figure 1), with a surface area of 1730 hectares and a mean depth of 11.2 m (maximum = 42.7 m). A thermocline is established in June-July and oxygen levels in the hypolimnion may drop to 1 to 10% of saturation.

Historical data on sport fish indicate that by 1978 the lake had become contaminated with a significant quantity of PCBs; probably Aroclor 1254. Contamination originated at the western edge of the lake (Figure 1), where PCB-laden oil was used on the road to suppress dust. In 1983, the contaminated road surface was removed to a containment facility. It has been estimated [Ontario Ministry of the Environment (OME), unpublished report], that the period between contamination and removal of the road surface was seven to eight years.

Historical data from 1978 to 1985 on concentrations of PCBs in Lake Clear fish and sediments were obtained from the Ontario Ministry of the Environment

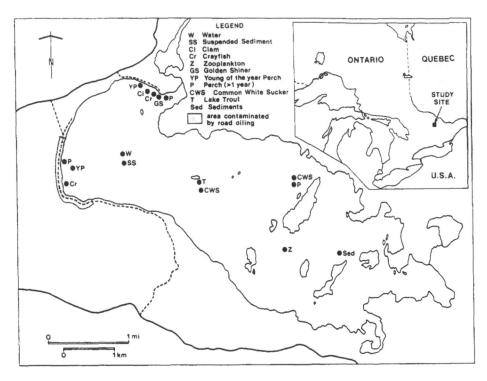

Figure 1.   Map of study area showing sampling sites and the location of Lake Clear in Ontario, Canada (inset).

(Table 1). These unpublished data were generated from the ministry's Ontario-wide monitoring program for sport fish, and from several specific OME projects on Lake Clear during 1981 to 1985. Fish were analyzed for total PCB residue levels on a wet weight basis and sediments were analyzed for total PCBs on a dry weight basis (μg/kg) by packed column gas chromatography (LRGC-ECD). Dorsal muscle tissues were analyzed for all fish in the monitoring program, with the exception of yearling yellow perch, where PCBs were analyzed from whole fish samples.

Sample Collection

Biota and sediments were collected in October 1986 and June 1987 in the areas shown in Figure 1. Details of biota collection and sample preparation are given elsewhere (Macdonald and Metcalfe 1989, 1991). Crayfish (*Orconectes* sp.), clams (*Elliptio complanata*), and "mixed invertebrates" (mainly trichopteran, ephemeropteran, and odonate insect larvae) were collected using hand-held dip-nets. Zooplankton were collected using a conical net made with 276 μm nytex mesh. Samples from two or three tows (five minutes each) were

Table 1. Summary of Historical Data of Total PCB Concentrations in Lake Clear Fish Determined by the Ontario Ministry of the Environment (Unpublished Data) from 1978 to 1985 Using Low Resolution GC. Values are Means with Ranges (in Brackets).

Species	Sample size (n)	Mean weight (g)	Mean length (cm)	% lipid	PCB concentration (µg/kg)
1978					
Lake trout	1	1092	48.3	2.7	2530
Northern pike	4	1316 (756–1960)	60.3 (50.8–71.0)	0.5	2197 (1300–3700)
1980					
Lake trout	5	2710 (2520–2912)	65.0 (62.4–67.2)	3.8	6319 (3564–9081)
Northern pike	16	1275 (252–4088)	56.1 (37.1–83.0)	0.4	353 (108–951)
1981					
Lake trout	18	3147 (2000–5500)	65.2 (56.0–75.0)	5.5	8253 (3092–15800)
Northern pike	20	1199 (500–2000)	56.5 (38.7–68.0)	0.1	406 (43–1623)
Smallmouth bass	1	300	28.0	1.0	727
Whitefish	10	923 (425–1600)	42.9 (37.0–49.0)	ND	2194 (326–6392)
Yellow perch	15	201 (150–300)	26.3 (24.6–28.2)	ND	321 (188–619)
Yellow perch (yearling)	5	NR	6.1 (6.0–6.3)	ND	1716 (1524–1968)

Table 1.   (continued)

Species	Sample size (n)	Mean weight (g)	Mean length (cm)	% lipid	PCB concentration (µg/kg)
1982					
Lake trout	12	3582 (1150–6015)	68.6 (49.0–83.5)	6.5	5933 (660–9100)
Lake trout	7	3296 (2240–4032)	66.2 (60.5–70.4)	6.1	6293 (1679–9397)
Northern pike	36	1255 (364–3248)	57.8 (31.2–82.2)	0.5	606 (128–1997)
Common white sucker	20	1308 (364–2128)	46.6 (30.8–56.6)	1.0	1105 (57–6376)
Smallmouth bass	11	873 (140–1568)	35.6 (22.5–45.1)	1.0	1194 (463–2258)
Burbot	21	1310 (182–4500)	50.4 (29.0–75.6)	0.3	305 (92–958)
Whitefish	20	1312 (378–2812)	48.9 (34.4–58.2)	3.6	5134 (559–12300)
Yellow perch	20	211 (168–294)	27.6 (25.8–29.6)	0.5	567 (250–1144)
Rock bass	20	343 (210–434)	25.2 (21.1–27.2)	0.5	760 (136–3849)

Note: NR - not reported; ND - not detected.

grouped to give a single sample and were further concentrated by passing the sample through 64 μm mesh.

Golden shiner (*Notemigonus crysoleucas* Mitchill) and adult and young-of-the-year (YOY) yellow perch (*Perca flavescens* Mitchill) were collected using a 20 m bag seine. White sucker (*Catostomus commersoni* Lacépède), large adult perch and a single lake trout (*Salvelinus namaycush* Walbaum) were collected in 1.3 or 2.6 m trap nets in October 1986. Except for the trap nets, all sampling equipment was soaked in reagent grade methanol prior to sampling. Whitefish (*Coregonus clupeaformis* Mitchill) and smallmouth bass (*Micropterus dolomieui* Lacépède) were collected by the Ontario Ministry of Natural Resources in Pembroke, Ontario. Total body weight and length of all fish were measured in the field prior to dissection and freezing. A sample of muscle tissue was taken from the dorsal area of the fillet below the dorsal fin for lake trout, bass, whitefish, white sucker, and adult yellow perch samples. With smaller fish (golden shiner, YOY perch), the head, tail, and internal organs were removed before analysis. Crayfish samples consisted of muscle dissected from the tail, and clam samples consisted of soft tissue removed from the valves. Since only one lake trout was collected, four subsamples of dorsal fillet were prepared for residue analysis of this fish.

Sediment cores were collected at a depth of approximately 30 m using a K-B corer with 0.35 m brass tubes. Three-centimeter sections of the core were removed by elevating a plug inside the tube and removing the sections by a solvent-rinsed aluminum sectioning device. Sediments for analysis of total organic carbon were collected by Eckman grab.

Water samples were collected by hand pumping water with a stainless steel and teflon solvent pump from a depth of approximately 1 m into a stainless steel container. Eighteen liters of lake water were passed through 137 mm diameter organic binder-free filters (pore size of 0.3 μm) into a 20-liter stainless steel extraction vessel and extracted with two volumes (500 and 400 ml) of distilled-in-glass methylene chloride. The 0.3 μm filters were stored in solvent-washed 250-ml Mason jars with approximately 100 ml of methylene chloride. Grab samples of water were collected for analysis of pH, alkalinity, suspended solids, and DOC. All sediment, suspended sediment, and biota samples were stored temporarily in solvent-washed Mason jars, which were sealed with solvent-washed aluminum foil. All biota samples were transferred to solvent-washed aluminum foil and frozen within 8 hr of collection.

Sample Preparation

Sediment samples of approximately 5 g were prepared for analysis using method "A" of the Ontario Ministry of the Environment protocol for the analysis of organochlorine pesticides and PCBs in sediments (Ontario Ministry of the Environment 1983). Sediments were extracted with acetone by sonication, back extracted into methylene chloride and dried by passing through sodium sulfate.

Samples were cleaned up and fractionated into three subfractions by column chromatography (1 cm ID) on 5 g of activated silica gel (60 to 200 mesh). The column was eluted with 40 ml of hexane to yield Fraction A which included PCBs, DDE, aldrin, lindane, heptachlor, and mirex. Sulphur compounds were removed from Fraction A by precipitation with mercury.

Biota samples of approximately 5 g were ground with sodium sulfate in a mortar, and extracted into hexane with a soxhlet apparatus for 1 hr. Half of the extract was rotary evaporated to 2 ml and made up to 5 ml in 55:45 hexane:methylene chloride. Lipids were removed from this extract by gel permeation chromatography on a 3-cm (ID) × 28 cm column of Biobeads SX2. The first 90 ml of eluent, which contained lipid, was collected and evaporated to dryness for gravimetric determination of lipid content. The next 100 ml of eluent, which contained PCBs and organochlorine pesticides, was collected and cleaned up using silica gel, as described previously.

The methylene chloride collected in the field from two water samples (2 × 18 l) was passed through sodium sulfate, pooled, and rotary evaporated to 2 ml. The methylene chloride was replaced by successive additions of hexane and evaporation under nitrogen. The sample was cleaned up and fractionated by silica gel chromatography, as described above. The filters containing the suspended particulate sample from two 18-liter water samples were pooled and manually broken up in the methylene chloride originally added in the field. The entire sample was filtered under vacuum through sodium sulfate in a Buchner funnel and a further 150 ml of methylene chloride passed through the filter. The extract was partitioned into hexane and cleaned up/fractionated by silica gel chromatography, as described previously.

Sample Analysis

Sediment samples were analyzed for organic carbon content and particle size distribution, and water samples were analyzed for pH, alkalinity, suspended solids, and DOC according to the protocols of the Ontario Ministry of the Environment (1983).

Samples were analyzed for PCBs by high resolution gas chromatography (HRGC-ECD) using the GC conditions described in Macdonald and Metcalfe (1989). Individual congeners were identified by retention time (window 0.02%) and quantified from integrated peak areas. The "limit of detection" (LOD) was calculated as three times the standard deviation of mean concentrations of each PCB congener detected in sample blanks (Keith et al. 1983). The limits of detection for the 19 congeners were between 0.3 and 1.0 ng/ml.

Quantification of individual PCB congeners was made by comparison to standards purchased from the National Research Council (NRC) in Halifax, Canada. The standard was a mixture of the NRC standards CLB-1 A and D, to which congener 52 (2,2′,5,5′) was added. DDE was added as a relative retention

Table 2.   Summary of IUPAC Numbers (Ballschmiter and Zell 1980), Log Octanol-Water Partition Coefficients (K_{ow}), and Chlorine Substitution of the PCB Congeners Analyzed in the Lake Clear Samples

Congener number	Chlorine number	log K_{ow}	Chlorine position
18	3	5.24	2,2′,5
31 (28)	3 (3)	5.67 (5.67)	2,4′,5 (2,4,4′)
52	4	5.84	2,2′,5,5′
49	4	5.85	2,2′,4,5′
44	4	5.75	2,2′,3,5′
101	5	6.38	2,2′,4,5,5′
87	5	6.29	2,2′,3,4,5′
110 (77)	5 (4)	6.48 (6.36)	2,3,3′,4′,6 (3,3′,4,4′)
151	6	6.64	2,2′,3,5,5′,6
118 (149)	5 (6)	6.74 (6.67)	2,3′,4,4′,5 (2,2′,3,4,5′,6)
153 (132)	6 (6)	6.92 (6.58)	2,2′,4,4′,5,5′ (2,2′,3,3′,4,6′)
138	6	6.83	2,2′,3,4,4′,5′
180	7	7.36	2,2′,3,4,4′,5,5′
170	7	7.27	2,2′,3,3′,4,4′,5
201	8	7.62	2,2′,3,3′,4,5,5′,6′
196	8	7.65	2,2′,3,3′,4,4′,5,6′
195	8	7.56	2,2′,3,3′,4,4′,5,6
194	8	7.80	2,2′,3,3′,4,4′,5,5′
209	10	8.18	2,2′,3,3′,4,4′,5,5′,6,6′

Note: Numbers in brackets refer to congeners co-eluting with major congeners. Log K_{ow} values are from Hawker and Connell (1988).

time marker. Nineteen congeners were chosen for analysis (Table 2), the sum of which presented approximately 46%, 53%, and 62% of the total PCBs present in Aroclors 1242, 1254, and 1260, respectively (Murphy et al. 1987, Manchester-Neesvig and Andren 1989, Schulz et al. 1989). Of the 19 congeners quantified from the standard, four of the congeners coeluted with other PCB congeners present in the environmental samples (Table 2). All coeluting peaks were reported as the concentration of the PCB congener present in the standard, with the exception of PCB 77 which coeluted with PCB 110. Since PCB 77 (3,3′,4,4,′) makes up a very small proportion of PCBs in commercial Aroclors in comparison to PCB congener 110 (2,3,3',4',6) (Schulz et al. 1989), it was assumed that the contribution by PCB 77 was negligible relative to PCB 110 in the Lake Clear samples. However, since there were no purified standards of PCB 110 available, PCB 77 in the standard was used as a retention time marker for PCB 110 and was also used to calculate the concentration of PCB 110 in Lake Clear samples. The integrated area of the peak eluting at the retention time for PCB 77 was multiplied by the response factor of PCB 77 times the ratio of the relative response factors for PCBs 77 and 110, as determined by Mullin et al. (1984). Peak identities and congener concentrations were confirmed by GC/MS of the selected samples.

Data Treatment and Statistical Analysis

PCB levels determined by HRGC-ECD are reported in this paper in two ways; either as the sum of the 19 congeners analyzed (i.e., total congener concentration or total PCB), or as the proportion of each congener relative to the sum of all individual congeners. Total congener concentrations in biota were calculated both on a wet weight basis and a lipid weight basis. Lipid levels in the 4 to 5 g samples were often close to detection limits for gravimetric analysis. Hence, for one clam and one whitefish sample the lipid value was estimated from the mean of the other four samples.

To relate historical PCB residue data to more recent data, it was assumed that the sum of the 19 PCB congeners analyzed by HRCG-ECD represented 50% of the total PCBs analyzed by LRGC-ECD. Thus, estimates of total PCB residues were made by doubling the total congener concentrations. Whole body concentrations of PCBs in fish were estimated by multiplying muscle concentrations by a factor of 2.5 (Niimi and Oliver 1983).

All data were analyzed using the personal computer version of SPSS software (SPSS/PC, Inc. version 3.0, Chicago, IL 1988). Significant differences were measured at the 0.05 level (i.e., $P \le 0.05$). All means and range tests were calculated using log transformed data to normalize the residuals.

Principal Component Analysis

Principal component analysis was conducted using SPSS/PC and EINSIGHT Ver. 2.1 (Infometrix, Inc., Seattle, Washington). The latter program uses methods of calculation common to chemometrics (Sharaf et al. 1986) and was used to support the analysis from SPSS/PC. The congener composition data for each sample were entered in a single matrix and rotated using a varimax rotation (SPSS/PC 1988, Sharaf et al. 1986). For the principal components analysis, random numbers between zero and the detection limit were entered for all data in which congener residues were undetectable (Sharaf et al. 1986). In addition, congener 209 (decachlorobiphenyl) was removed from the principal component analysis because it was rarely detected in Lake Clear samples.

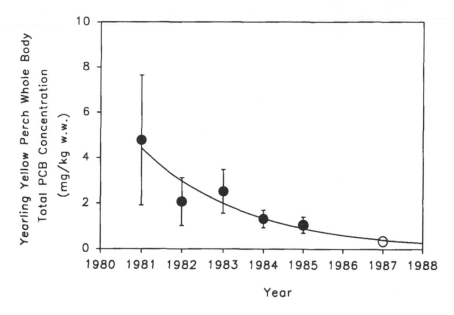

Figure 2. **Decline in the concentration of total PCB in the whole body of yearling yellow perch. Closed symbols are from the Ontario Ministry of the Environment (unpublished data), while the open circle is estimated from the young-of-the-year congener data of the present study. The solid line has been fitted to the data with a negative exponential equation (see text).**

Results

Historical Monitoring Data

By 1978, total PCB concentrations in the dorsal muscle of lake trout and northern pike (*Esox lucius*) in Lake Clear were greater than 2,000 µg/kg wet weight (Table 1). PCB residues from lake trout collected in 1980, 1981, and 1982 ranged widely from 660 µg/kg to 15,800 µg/kg. Whitefish, smallmouth bass, adult yellow perch, and common white suckers were also heavily contaminated with PCBs. A survey of whole body PCB residues in yearling yellow perch collected during 1981 to 1985 from an area in the western portion of the lake which is adjacent to the contaminated area indicated that there was a decline in mean total PCB concentrations from a maximum of 4770 µg/kg in 1981 to 907 µg/kg in 1985 (Figure 2).

Samples of the contaminated road surface and adjacent lake sediments collected in 1981 and 1982 revealed very high concentrations of total PCBs. The road surface contained up to 300 mg/kg PCBs (dry weight), and lake sediments 15 to 30 m offshore contained from 1 to 10 mg/kg of total PCBs (Ontario Ministry of the Environment, unpublished data). In 1982, PCB concentrations in surface sediments from the deepest area of the lake (see area marked "sed" in Figure 1) were 0.45 mg/kg dry weight.

Total Congener Concentration
Water, Suspended Solids, and Sediments

Water samples collected in 1987 had a dissolved organic carbon (DOC) concentration of 2.83 mg/l (SD = 0.05, n = 6), alkalinity of 111.2 mg/l (SD = 0.34, n = 6) and pH of 8.4. Suspended solids, which were composed primarily of detritus and phytoplankton, were at a concentration of 1.08 mg/l (SD = 0.78, n = 10). The concentration of total PCB congeners dissolved in water as 1.95 ng/l (n = 2), and the concentration on suspended solids was 897 μg/kg dry weight (n = 2).

Size distribution analysis of the sediments indicated that 65% (SD = 5.1, n = 5) of the particles were less than 45 μm, with the remaining 35% in the range of 45 to 1000 μm. As indicated in Table 3 for samples collected in 1987, the mean concentration of total PCB congeners in the surface sediment (0 to 3 cm) was 597 μg/kg dry weight, but the concentration decreased with depth. Total organic carbon content of the sediment was 11.2% (SD = 0.13, n = 5), giving a concentration of 5.35 mg/kg carbon in the 0 to 3 cm section (Table 3).

Biota

For biota samples collected in 1986–87, the geometric means and standard deviations for total body weight, total length, lipid content, and total congener concentrations on both a wet weight and lipid weight basis are summarized in Table 3. Mean ages of mature fish ranged from approximately 4 years in smallmouth bass and adult perch (as determined by scale reading), to approximately 12 years in the single lake trout (as estimated from age-weight regression for Ontario lakes in Hoyle and Stewart, 1987). Yellow perch spanned a range of ages from zero to seven, with the intermediate sized fish estimated to be approximately one and two years. Ages of the white suckers and whitefish were estimated from body length to be approximately ten years and four years, respectively, based on data from similar lakes (Scott and Crossman 1973).

Lipid values ranged from a minimum of 0.15% in the whole body sample of the clams to 11.3% in the muscle tissue of golden shiner. Golden shiners, the single lake trout, and zooplankton contained the highest levels of lipid. Excluding the lake trout, there was no correlation between the lipid content of the biota and total PCB concentrations on a wet weight basis (P > 0.05). Total PCBs (wet weight) ranged from 23.6 μg/kg in the clam sample to 2993 μg/kg in the lake trout. However, the majority of biota samples were in the range of 60 to 150 μg/kg. Excluding the lake trout sample, the highest wet weight PCB residue levels were in smallmouth bass, white suckers, and whitefish. Scheffé's multiple range test grouped these species with YOY perch and golden shiner as having significantly greater PCB concentrations than clams and zooplankton.

There were some shifts in the relative concentrations of total congeners in biota when expressed on a lipid weight basis (Table 3). Maximum PCB concentrations were observed in lake trout, crayfish, and older perch (i.e., greater than one year). Smallmouth bass and white sucker, which contained among the highest residue concentrations on a wet weight basis, were in the middle of the range on a lipid basis.

Yellow perch was the only species for which sufficient samples were analyzed to permit statistical tests of relationships between PCB levels and age. There was no significant relationship between PCB concentration and age on either a wet weight or lipid weight basis, even though muscle lipid levels decreased significantly with age (P < 0.001, $R^2 = 0.53$). As another approach, whole body concentrations of total congeners were estimated as 2.5 times the muscle concentration, and whole body burdens calculated as [estimated body concentration × body weight]. There was a strong relationship between age and the log of total body burden in yellow perch (P < 0.001, $R^2 = 0.88$), with a slope of 0.306.

Congener Distribution

The concentration of the individual congeners in sediments, water, suspended solids, and biota are shown in Figure 3 as a proportion of the total PCB congeners. The major congeners in all biota samples, including zooplankton, are the pentachlorobiphenyl compounds 87, 101, 110 and 118 and the hexachlorobiphenyl congeners 153 and 138; the same congeners that are dominant in Aroclor 1254. These same congeners are prevalent in all three sections of the sediment core, although only the data for surface sediment (0 to 3 cm) and the lowest segment (6 to 9 cm) are shown in Figure 3. The congener patterns in water and suspended solids are different from all biota and sediment samples, and showed a higher proportion of less chlorinated congeners.

Shifts in the proportion of individual congeners analyzed in biota, sediment, water, and suspended sediments were examined statistically using two methods. The first method consisted of analyzing the proportions of five structurally related congeners (PCBs 18, 52, 101, 180 and 194) in water, sediment, and biota, using analysis of variance and range tests to determine significant differences between groups (Figure 4). These congeners are similar in that they have chlorine substitution at the 2,2′,5 positions on the biphenyl rings. This analysis, which is similar to that of Oliver and Niimi (1988), indicates that the proportion of a tri-(PCB 18) and tetrachlorobiphenyl (PCB 52) to total congeners was significantly higher in water, suspended solids, and invertebrates, while the proportion of a hexachlorobiphenyl (PCB 153) was elevated in biota from upper trophic levels. The most highly chlorinated congener (PCB 194) remained at a constant proportion throughout all compartments (Figure 4).

Table 3.   Geometric Means and Standard Deviations (in Brackets) of Body Weight, Total Length, PCB Concentration, and Lipid Content of Lake Clear Biota

Sample group	Sample size (n)	Body Weight (g)	Total Length (cm)	% Lipid	PCB concentration	
					wet weight (µg/kg)	lipid weight (mg/kg)
Benthos						
Crayfish	5	20.9 (11.2–39.0)	8.04 (6.67–9.69)	0.19 (0.10–0.35)	7.32 (38.0–141)	39.1 (11.8–130)
Clams	5	44.7 (39.1–51.2)	8.12 (7.58–8.70)	0.15 (0.06–0.42)	23.6 (15.5–36.0)	14.5 (8.71–24.1)
Miscellaneous invertebrates	1	–	–	0.58	34.7	3.0
Zooplankton						
Pooled sample	5	–	–	1.17 (1.00–1.38)	59.3 (47.7–73.9)	5.06 (4.51–5.68)
Fish						
Golden shiner	5	1.73 (1.27–2.36)	6.18 (5.32–7.19)	11.3 (6.55–19.4)	79.7 (69.9–91.0)	0.71 (0.38–1.33)
Perch (>1y)	15	67.7 (22.5–204)	19.0 (13.6–26.6)	0.22 (0.17–0.28)	76.3 (47.7–122)	35.1 (21.3–57.7)
Percy (YOY)	5	2.74	6.89	0.48	89.8	18.9

	n					
Perch (>1 yr)	15	67.7 (22.5–204)	19.0 (13.6–26.6)	0.22 (0.17–0.28)	76.3 (47.7–122)	35.1 (21.3–57.7)
Perch (YOY)	5	2.74 (1.94–3.87)	6.89 (6.19–7.67)	0.48 (0.36–0.64)	89.8 (79.3–102)	18.9 (13.0–27.4)
Common white sucker	5	955	51.5	0.49	121	24.5
Smallmouth bass	5	720 (601–1520)	35.5 (47.1–56.3)	0.75 (0.30–0.81)	153 (63.2–232)	20.2 (18.7–32.2)
Whitefish	5	627 (682–760)	39.9 (33.9–37.2)	0.39 (0.59–0.97)	104 (125–186)	26.6 (16.7–24.5)
Lake trout	1	2100 (453–867)	76.6 (36.9–43.2)	6.80 (0.26–0.57)	2993 (57.9–188)	46.8 (17.8–39.7)
Sediment						
0–3 cm (surface)	5	—	—	11.2	571 (410–796)	5.35 (4.19–6.82)
3–6 cm	5	—	—	11.2	357 (300–424)	3.19 (2.69–3.79)
6–9 cm	4	—	—	11.2	242 (183–320)	2.16 (1.63–2.86)

Note: Sample size (n) refers to the number of samples analyzed for each parameter. Body weight of clam and crayfish samples are based on four individual samples as are the lipid values for the clam and whitefish samples. Value for sediment in lipid column is total organic carbon (TOC). PCB concentrations reported for sediment are concentrations on a dry weight and on a total carbon basis in respective columns.

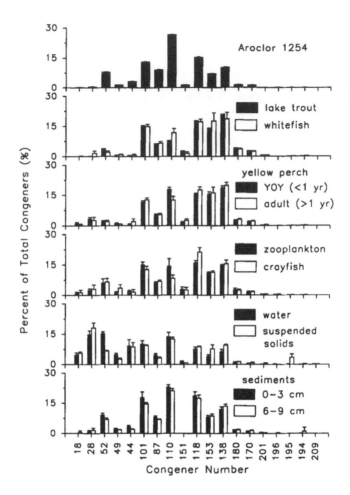

Figure 3. The proportion of individual congeners in water, suspended solids, surface sediments and six groups of biota as a percent of the total congeners. Each bar in the histograms represents a mean value (the number of samples is given in Table 3) with error bars of one SD. The histogram for Aroclor 1254 is the mean of five determinations.

In the second type of analysis, principal component procedures were used to illustrate the degree of similarity or dissimilarity in congener patterns among the different compartments of the Lake Clear ecosystem. Five determinations of each of Aroclors 1242, 1248, 1254 and 1260 standards were included in the analysis as reference points and to indicate the sensitivity of the analysis to changes in the distribution of congeners. Principal component analysis of all sediment, water, suspended solids, biota, and the standards reduced the data to three factors which accounted for 85% of the total variability. The first principal

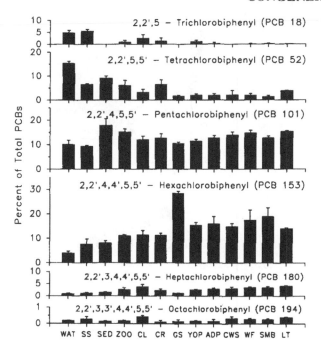

Figure 4. Comparison of six structurally similar congeners (2,2',5 substitution) in all samples from Lake Clear. Legend: WAT, water; SS, suspended solids; SED, surface sediments; ZOO, zooplankton; CL, clams; CR, crayfish; GS, golden shiner; YOP, young-of-the-year yellow perch; ADP, adult yellow perch (>1 year); CWS, common white sucker; WF, whitefish; SMB, smallmouth bass; LT, lake trout.

component, (40.9% of the variability), consisted of all the hepta- and octachlorobiphenyls (PCBs 196, 201, 180, 170, 194, 195) plus the hexachlorobiphenyl congener 151, and separated Aroclor 1260 from all other groups (Figure 5). There was no similarity between any of the samples and Aroclor 1260 (Figure 5). The second principal component, which consisted of tetrachloro- and hexachlorobiphenyls (PCBs 52, 44, 153, 49 and 138) produced the greatest separation between samples and Aroclor mixtures (Figure 5), and was responsible for 35% of the total variation. Highest positive loadings of the second factor were for PCBs 153 and 138, while the highest negative loadings were from PCBs 52, 44 and 49. When the data were reanalyzed without Aroclor 1260, the second component became dominant, accounting for 50% of the variability. The third component of the full analysis explained 9.0% of the variability and consisted of tri- and pentachlorobiphenyls.

Separation of the Lake Clear samples was most evident along the second principal component (tetra- and hexachlorobiphenyls), with the water samples

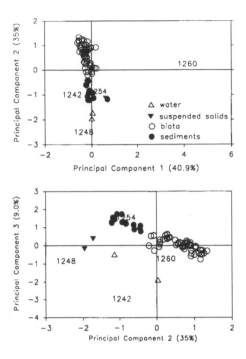

Figure 5.   Principal component analysis of congener proportions in water, suspended solids, sediments and biota showing the three major components that account for 85% of the variability of the data. The numbers on the graphs refer to five determinations of Aroclor standards.

situated midway between Aroclors 1248 and 1254. Suspended solids samples were variable, and were not aligned with any Aroclor mixture with respect to the second component. These groupings reflect the higher proportion of less chlorinated congeners in water and suspended solids. All bottom sediment samples showed a strong similarity with Aroclor 1254 in all three principal components, indicating that the full range of congeners was the same in sediment and the Aroclor. In general, all biota responded to the principal components analysis as a group and there was no consistent differentiation of any particular species or trophic level. Hence, in Figure 5, all biota are represented by a single symbol. Sediments were closely associated with Aroclor 1254 in the third component by high loadings on PCBs 101, 87, and 110.

Discussion

The historical data indicate that a significant quantity of PCBs, probably Aroclor 1254, entered Lake Clear in the mid-1970s, although the exact time of contamination is difficult to determine. Data from 1986–87 indicate that contamination is still present in all groups of biota, water and sediment, although the concentrations are generally lower than in the early 1980s. For example, total PCB levels in whitefish and smallmouth bass decreased from mean values of 5134 and 1194 µg/kg in 1982 (Table 1) to approximately 208 and 306 µg/kg, respectively, in 1987, assuming that the 19 congeners analyzed in the present study account for 50% of total PCBs. PCB concentrations in lake trout appear to have remained relatively constant from 1980 to 1987 (approximately 6,000 µg/kg), although data interpretation was hampered by the extreme variability in concentration between fish and the small sample size in 1987 (n = 1). In addition, the age of the fish collected in 1987 indicates that it was probably present in the lake at the time of the original spill and has been exposed to the contaminated conditions for ten + years.

A decrease in total PCB concentrations is most evident in yearling yellow perch samples, where concentrations declined from a maximum mean of 4770 µg/kg in 1981 to 907 µg/kg in 1985 in the area of contamination. When total congener data for muscle tissue is converted to whole body total PCBs, the levels in YOY perch collected in the present study in 1987 are equivalent to approximately 350 µg/kg; consistent with a continuing decline in PCB levels (Figure 2). An exponential decay relationship fitted to the yearling/YOY perch data (Figure 2) corresponds to a half-life of 1.73 years.

Declining PCB concentrations in biota suggest that the amount of PCB available to the biota has decreased at a significant rate. Possible major routes of loss of PCBs include volatilization (Swackhamer et al. 1988), microbial degradation (Brown et al. 1987, Quensen et al. 1988), and adsorption onto particulates followed by sedimentation (Thomann et al. 1987).

Although Henry's Law constants for individual PCB congeners cover a broad range of values (Burkhard et al. 1985a, Shiu and Mackay 1986, Murphy et al. 1987, Dunnivant et al. 1988), volatilization of PCBs has been suggested as a major mechanism for the removal of PCBs from lake systems, especially among less chlorinated PCB congeners. Swackhamer et al. (1988) constructed a mass balance for an isolated lake which showed that the losses of PCBs to the atmosphere were approximately equal to the losses by sedimentation, and that the highest volatilization losses were for PCBs 18 (trichlorobiphenyl), 52 and 49 (tetrachlorobiphenyls), and 101 (pentachlorobiphenyl). In the present study, it would be expected that these congeners, plus PCBs 31(28) and 44, would be most susceptible to losses through volatilization.

Similarly, microbial degradation of lower chlorinated PCB congeners does occur in sediments (Brown et al. 1987, Quensen et al. 1988), although the extent of the process is unclear (Brown et al. 1988) and appears to only occur at very high concentrations of PCB. Although primarily trichlorobiphenyls and tetrachlorobiphenyls are degraded in pure culture (Furukawa et al. 1978), in natural sediments, the position of chlorines on the biphenyl ring can be a strong determinant of degradability. Brown et al. (1987) reported microbial dechlorination of congeners 18, 52, 49, 110, and 118.

The distribution of major PCB congeners in the samples collected in Lake Clear in 1986 and 1987 indicate that losses due to either volatilization or microbial degradation are undetectable at this point. The presence of less chlorinated congeners in the water phase suggests that there is still a large pool of tri- and tetrachlorobiphenyls in the lake. Using the values of Swackhamer et al. (1988) as conservative estimates of the quantity of congeners lost to the atmosphere from a small northern lake (710 ng/m/year), the approximately 12 g per year of PCB 18 alone would volatilize from Lake Clear, and about half that amount of PCBs 101 and 52 would be lost. Assuming that the replacement of these congeners through atmospheric sources is low compared to the original input (Macdonald and Metcalfe 1991), we would have expected a skewing of congener distribution towards more highly chlorinated compounds (Burkhard et al. 1985b), had volatilization been a major fate process.

The same argument applies for the possible losses of PCBs through microbial degradation. The lower chlorinated congeners, such as PCBs 52 and 49, which appear to be particularly susceptible to dechlorination in sediments (Brown et al. 1987), remain at a relatively constant proportion of the total congeners at all depths in the sediment core. Principal component analysis grouped all sediment samples, including those taken 6 to 9 cm below the surface, with Aroclor 1254; indicating a strong similarity between the PCBs in sediment and the Aroclor with all three factors.

Further evidence for the lack of weathering of the PCBs can be seen in the congener distribution within tissues of short-lived organisms, such as zooplankton and YOY perch, which have been exposed to Lake Clear contamination for only a matter of months. The samples are generally lower in total congener concentration than the higher trophic levels, but demonstrate the same congener patterns as fish which have been in the lake for 10 to 12 years (whitefish and the lake trout).

This comparison of the historic data with the concentrations of PCBs in 1986 to 1987 indicates that considerable quantities of PCBs have been removed from general circulation within the lake, and comparison of congener patterns shows that this removal is occurring at a relatively constant rate among all PCB congeners. Since volatilization and microbial degradation seem to be relatively minor processes, the general decline in the PCBs in the water column is probably occurring by partitioning into sediments. Sedimentation of PCBs in Lake Clear appears to be an ongoing process since, in 1987, the highest total congener concentrations in the sediment were 597 µg/kg in the top 3 cm (corresponding to 1194 µg/kg total PCB), and the pattern of congeners was constant from the surface to 9 cm. In 1982, total PCB concentrations in the surface sediments at this same site were 450 µg/kg. The predominant flow pattern within the lake is probably from the shallow, contaminated west end to the deposition zone at the eastern end, since the principal winds are from the northwest. In 1987, total congener concentrations in the sedimenting material (suspended solids) were higher (897 µg/kg) than in the surface sediment. These data suggest that PCBs were still being deposited at a significant rate in 1986–1987, and that adsorption onto suspended solids, followed by transport to the deposition zone at the eastern end of the lake, was the dominant removal process. This is in contrast to atmospherically contaminated lakes where volatilization and sedimentation probably both account for a significant proportion of the losses of PCBs from the lakes (Swackhamer et al. 1988, Macdonald and Metcalfe 1991). Other factors such as sediment organic content as well as lake specific characteristics such as lake morphometry, amount of ice cover during the year and temperature regimen of the surface waters, probably also affect the specific PCB congeners present in a lake and hence the relative importance of the individual control mechanisms (Macdonald and Metcalfe 1989, 1991).

Using simulations based on fugacity models, Burkhard et al. (1985a) predicted that there will be little weathering of PCBs in aquatic systems, unless the various compartments within the system experience multiple equilibrations. A model based on the latter scenario which simulated the continuous movement of "clean" air across PCB-contaminated water indicated that less chlorinated congeners (i.e., highest Henry's Law constants) would be preferentially removed; resulting in an enrichment of more highly chlorinated congeners with time in all compartments of the system. In "sedimentation" simulations, the continuous renewal of "clean" suspended particulate material to the system resulted in an initial enrichment of more highly chlorinated congeners (i.e., highest K_p values) in bottom sediments, but this trend decreased with time. There is no evidence that either of these processes are active in Lake Clear. Since enrichment of highly chlorinated congeners has been noted in the Hudson River sediments (Brown et al. 1985), the lack of weathering noted in Lake Clear may be a specific characteristic of contaminated cold water lake ecosystems.

The temporal trends in PCB residue concentrations within biota have significant implications to the sport fish industry. Within Lake Clear biota, there appeared to be no differences in PCB concentration between pelagic and benthic organisms, or between trophic levels, when compared on a lipid basis. Surprisingly, the highest concentrations on a lipid weight basis were observed in crayfish muscle even though clams collected from the same area were low in total PCB. There was no significant difference between PCB residues on a lipid weight basis in top predators (smallmouth bass, lake trout) versus omnivorous fish (white sucker, perch), contrary to other studies where yellow perch showed significantly higher concentrations of total PCB residues when compared against lower trophic levels (Macdonald and Metcalfe 1991).

The single factor which does appear to influence PCB concentration is time of exposure, as indicated by the age of the organism. Although it is not statistically significant, PCB concentrations (lipid basis) in short lived organisms (YOY perch, zooplankton) contain low PCB concentrations relative to lake trout, smallmouth bass, whitefish, and white sucker. These observations are consistent with the conclusions of Weininger (1978) and van der Oost et al. (1988), who concluded that the concentration of PCBs in lake trout in Lake Michigan are better described as a function of the age of the fish and length of time of exposure than the amount of lipid or the partition coefficient of the PCB mixture.

Accumulation of PCBs with age was observed in yellow perch from Lake Clear by an increase in total body burden. Regressions relating perch age to total body weight and estimated body burden of PCBs give the equations:

$$\log \text{ (body weight, g)} = 0.582 + 0.322 \text{ age(yr)} \quad R^2 = 0.88 \quad (1)$$

$$\log \text{ (body burden, μg)} = -0.076 + 0.306 \text{ age(yr)} \quad R^2 = 0.85 \quad (2)$$

Hence, between the first and second year, both the body weight and body burden of PCBs in perch increase at about the same rate, i.e., by a factor of approximately two. If it is assumed that the fish feed on a diet of zooplankton with a PCB concentration of 60 μg/kg, at a feeding rate of 2% body weight per day, with a PCB uptake efficiency of 60% (Thomann et al. 1987), then the fish will accumulate approximately 2.5 μg/yr from its diet alone; close to the observed value of 1.7 μg. Once PCBs are accumulated in fish, the clearance rates for the individual congeners are low and are approximately equal to zero for some of the highly chlorinated congeners (Niimi and Oliver 1983). Thus, the declines in PCB concentrations noted in sport fish within Lake Clear may be taking place as older, more contaminated fish disappear from the population, or as growth of new tissue under less contaminated conditions causes dilution of the more highly contaminated tissues.

The distribution of individual PCB congeners in biota followed expected patterns, in that the proportions of higher chlorinated congeners were statistically higher in biota than in water and suspended solids. The proportion of some of the congeners increased significantly through the food web (e.g., PCB 153), but, in general, the congener patterns in all biota samples were remarkably consistent. Oliver and Niimi (1988) observed that the greatest difference in congener composition throughout the Lake Ontario food web was between plankton and the water phase. Similarly, in Lake Clear, there is a marked difference in dominant congeners between zooplankton and water, but a strong similarity within biota as a whole. This is shown in the distinct grouping of biota in the principal components analysis, but also in the similarity of congener distribution between zooplankton, YOY perch and the much older lake trout and whitefish. Hence, the enrichment of moderately chlorinated congeners by biota is the same for both "young" and "old" individuals, and does not reflect weathering of PCBs with time. Instead, it shows a relatively rapid equilibrium between biota and their environment for all congeners.

An examination of temporal changes and distribution of PCBs in Lake Clear shows that residue levels in biota have declined measurably within approximately ten years postcontamination. In contrast, within sediments in a deposition zone of the lake, PCB concentrations appear to have increased in recent years. The stable pattern of congeners in sediments and within biota of all ages suggests that PCB weathering has not occurred with time. These data indicate that volatilization is not a major route for the loss of less chlorinated PCB congeners from this contaminated lake. This may be a phenomenon which is specific for contaminated cold water lakes, as opposed to contaminated rivers, lakes receiving only atmospheric deposition or warm water lakes with open water for the whole year. If this is the case, Lake Clear may be a relatively simple model for predicting the fate of PCBs in larger contaminated lakes, such as Lake Ontario, Lake Michigan, and Lake Erie.

ACKNOWLEDGMENTS

We thank Maritsa Bailey for assistance in the analysis of samples, and Ted Wolkowski and Jan Campfens for the collection of samples. Historical data for Lake Clear were kindly supplied by A.F. Johnson and Glenn Owens of the OME in Toronto and Kingston, Ontario, respectively. Equipment and fish samples were also supplied by A. Armstrong and C. Buckingham of the Ministry of Natural Resources in Pembroke, Ontario. We would like to thank John Ralston and the Research Advisory Liaison Committee of the OME for their advice throughout the study. Funding for the project was supplied by the Ontario Ministry of the Environment. Chromatographic data analysis equipment was purchased through a grant from the Helen McCrea Peacock Foundation. GC/MS confirmation of sample analysis was conducted by M. Simon and M. Mulvihill of the Canadian Wildlife Service, Hull, Québec, Canada.

REFERENCES

Addison, R.F., P.F. Brodie and M.E. Zinck. "DDT has Declined More than PCBs in Eastern Canadian Seals during the 1970s," *Environ. Sci. Technol.* 18:935-937 (1984).

Ballschmiter, K., and M. Zell. "Analysis of Polychlorinated Biphenyls (PCB) by Glass Capillary Gas Chromatography," *Fresenius Z. Anal. Chem.* 302:20-31 (1980).

Brown, J.F., Jr., D.L. Bedard, M.J. Brennan, J.C. Carnahan, H. Feng, and R.E. Wagner. "Polychlorinated Biphenyl Dechlorination in Aquatic Sediments," *Science* 236:709-712 (1987).

Brown, M.P., B. Bush, G.Y. Rhee and L. Shane. "PCB Dechlorination in Hudson River Sediment," *Science* 240:1674-1675 (1988).

Brown, M., M.B. Werner, R.J. Sloan and K.W. Simpson. "Polychlorinated Biphenyls in the Hudson River," *Environ. Sci. Technol.* 19:656-661 (1985).

Burkhard, L.P., D.E. Armstrong and A.W. Andren. "Henry's Law Constants for the Polychlorinated Biphenyls," *Environ. Sci. Technol.* 19:590-596 (1985a).

Burkhard, L.P., D.E. Armstrong and A.W. Andren. "Partitioning Behaviour of Polychlorinated Biphenyls," *Chemosphere* 14:1703-1716 (1985b).

Connell, D.W. "Bioaccumulation Behavior of Persistent Organic Chemicals with Aquatic Organisms," *Rev. Environ. Contam. Toxicol.* 101:117-154 (1988).

Devault, D.S. "Contaminants in Fish from Great Lakes Harbors and Tributary Mouths," *Arch. Environ. Contam. Toxicol.* 14:587-594 (1985).

Dunnivant, F.M., J.T. Coates and A.W. Eizerman. "Experimentally Determined Henry's Law Constants for 17 Polychlorobiphenyl Congeners," *Environ. Sci. Technol.* 22:448-453 (1988).

Furukawa, K., K. Tonomura and A. Kamibayashi. "Effect of Chlorine Substitution on the Biodegradability of Polychlorinated Biphenyls," *Appl. Environ. Microbiol.* 35:223-227 (1978).

Hawker, D.W., and D.W. Connell. "Octanol-Water Partition Coefficients of Polychlorinated Biphenyl Congeners," *Environ. Sci. Technol.* 22:382-387 (1988).

Hoyle, J.A., and T.J. Stewart. "Lake Trout Exploitation Stress Assessment on Drag, Dickey, and Twelve Mile Lakes," Ontario Ministry of Natural Resources, Bancroft, Ontario (1987), p. 50.

International Joint Commission. "1989 Report on Great Lakes Water Quality," Great Lakes Water Quality Board Report to the International Joint Commission, (1989), p. 128.

Kannan, N., S. Tanabe and R. Tatsukawa. "Potentially Hazardous Residues of Non-*Ortho* Chlorine Substituted Coplanar PCBs in Human Adipose Tissue," *Arch. Environ. Health* 43:11-14 (1988).

Keith, L.H., W. Crummett, J. Deegan, R.A. Libby, J.K. Taylor and G. Wentler. "Principles of Environmental Analysis," *Anal. Chem.* 55:2210-2218 (1983).

Kenaga, E.E., and C.A.I. Goring. "Relationship Between Water Solubility, Soil Sorption, Octanol-Water Partitioning, and Concentration of Chemicals in Biota," in *Aquatic Toxicology, Vol. 707*, J.G. Eaton, P.R. Parrish and A.C. Hendricks, Eds., (Philadelphia: ASTM, 1980), pp.78-115.

Macdonald, C.R., and C.D. Metcalfe. "A Comparison of PCB Congener Distributions in Two Point-Source Contaminated Lakes and One Uncontaminated Lake in Ontario," *J. Water Pollut. Control Fed. Can.* 24:23-46 (1989).

Macdonald, C.R., and C.D. Metcalfe. "Concentration and Distribution of PCB Congeners in Isolated Ontario Lakes Contaminated by Atmospheric Deposition," *Can. J. Fish. Aquat. Sci.* 48:371-381 (1991).

Manchester-Neesvig, J.B., and A.W. Andren. "Seasonal Variation in the Atmospheric Concentration of Polychlorinated Biphenyl Congeners," *Environ. Sci. Technol.* 23:1138-1148 (1989).

Mullin, M.D., C.M. Pochini, S. McCrindle, M. Romkes, S.H. Safe and L.M. Safe. "High Resolution PCB Analysis: Synthesis and Chromatographic Properties of all 209 PCB Congeners," *Environ. Sci. Technol.* 18:468-476 (1984).

Murphy, T.J., M.D. Mullin and J.A. Meyer. "Equilibration of Polychlorinated Biphenyls and Toxaphene with Air and Water," *Environ. Sci. Technol.* 21:115-162 (1987).

Niimi, A., and B. Oliver. "Biological Half-Lives of Polychlorinated Biphenyl (PCB) Congeners in Whole Fish and Muscle of Rainbow Trout (*Salmo gairdneri*)," *Can. J. Fish. Aquat. Sci.* 40:1388-1394 (1983).

Norstrom, R.J., M. Simon, D.C.G. Muir and R.E. Schweinsburg. "Organochlorine Contaminants in Arctic Marine Food Chains: Identification, Geographical Distribution, and Temporal Trends in Polar Bears," *Environ. Sci. Technol.* 22:1063-1071 (1988).

Oliver, B.G., and A.J. Niimi. "Trophodynamic Analysis of Polychlorinated Biphenyl Congeners and Other Chlorinated Hydrocarbons in the Lake Ontario Ecosystem," *Environ. Sci. Technol.* 22:388-397 (1988).

Ontario Ministry of the Environment. "Handbook of Analytical Methods for Environmental Samples," Laboratory Services and Applied Research Branch, Rexdale, Ontario, Canada (1983).

Quensen, III, J.F., J.M. Tiedje and S.A. Boyd. "Reductive Dechlorination of Polychlorinated Biphenyls by Anaerobic Microorganisms from Sediments," *Science* 242:752-754 (1988).

Sawhney, B.L. "Chemistry and Properties of PCBs in Relation to Environmental Impacts," in *PCBs and the Environment, Vol. 3*, Waid, J.S., Ed., (Boca Raton, FL: CRC Press, 1987), pp.47-64.

Schmitt, C.J., J.L. Zajicek and M.A. Ribick. "National Pesticide Monitoring Program: Residues of Organochlorine Chemicals in Freshwater Fish, 1980-81," *Arch. Environ. Contam. Toxicol.* 14:225-260 (1985).

Schulz, D.E., G. Petrick and J.C. Duinker. "Complete Characterization of Polychlorinated Biphenyl Congeners in Commercial Aroclor and Clophen Mixtures by Multidimensional Gas Chromatography-Electron Capture Detection," *Environ. Sci. Technol.* 23:852-859 (1989).

Scott, W.B., and E.J. Crossman. "Freshwater Fishes of Canada," *Fish. Res. Board Can.*, Bull. 184 (1973), p. 966.

Sharaf, M.A., D.L. Illman and B.R. Kowalski. *Chemometrics* (Toronto, Canada: John Wiley and Sons, 1986), p. 332.

Shiu, W.Y., and D. Mackay. "A Critical Review of Aqueous Solubilities, Vapor Pressures, Henry's Law Constants, and Octanol-Water Partition Coefficients of the Polychlorinated Biphenyls," *J. Phys. Chem. Ref. Data* 15:911-929 (1986).

Suns, K., G.G. Craig, G. Crawford, G.A. Rees, H. Tosine and J. Osborne. "Organochlorine Contaminant Residues in Spottail Shiners (*Notropis hudsonius*) from the Niagara River," *J. Great Lakes Res.* 9:335-340 (1983).

Suns, K., G.E. Crawford, D.D. Russell and R.E. Clement. "Temporal Trends and Spatial Distribution of Organochlorine and Mercury Residues in Great Lakes Spottail Shiners (1975-1983)," Ontario Ministry of the Environment (1985), p. 43.

Swackhamer, D.L., and D.E. Armstrong. "Estimation of the Atmospheric and Nonatmospheric Contributions and Losses of Polychlorinated Biphenyls for Lake Michigan on the Basis of Sediment Records of Remote Lakes," *Environ. Sci. Technol.* 20:879-883 (1986).

Swackhamer, D.L., B.D. McVeety and R.A. Hites. "Deposition and Evaporation of Polychlorobiphenyl Congeners to and from Siskiwit Lake, Isle Royale, Lake Superior," *Environ. Sci. Technol.* 22:664-672 (1988).

Thomann, R.V., J.P. Connolly and N.A. Thomas. "The Great Lakes Ecosystem-Modelling of the Fate of PCBs," in *PCBs and the Environment*, Waid, J.S., Ed., (Boca Raton, FL: CRC Press, 19877), pp.153-180.

van der Oost, R., H. Heida and A. Opperhuizen. "Polychlorinated Biphenyl Congeners in Sediments, Plankton, Molluscs, Crustaceans, and Eel in a Freshwater Lake: Implications of Using Reference Chemicals and Indicator Organisms in Bioaccumulation Studies," *Arch. Environ. Contam. Toxicol.* 17:721-729 (1988).

Weininger, D. "Accumulation of PCBs by Lake Trout in Lake Michigan," Ph.D. thesis, University of Wisconsin-Madison (1978), p. 232.

INDEX

R

S